Faster Than Light

Faster Than Light

Michael Gfoeller

Academica Press
Washington

Library of Congress Cataloging-in-Publication Data
Names: Gfoeller, Michael (author)
Title: Faster than light | Gfoeller, Michael.
Description: Washington : Academica Press, 2026.
Identifiers: ISBN 9781680536010 (hardcover) | ISBN 9781680536027 (ebook)

Library of Congress registration information for this title is in progress.

I offer my deep appreciation to my beloved wife, Tatiana Gfoeller, and my comrade and editor, Kurt Klein. We have already shared epic journeys through the corridors of time, and there is much more to come!

This book is dedicated to Eratosthenes of Cyrene, head of the Library of Alexandria, the first mathematician to calculate the true size and shape of the Earth. Some 1687 years after his death, his work led Christopher Columbus to the discovery of a New World.

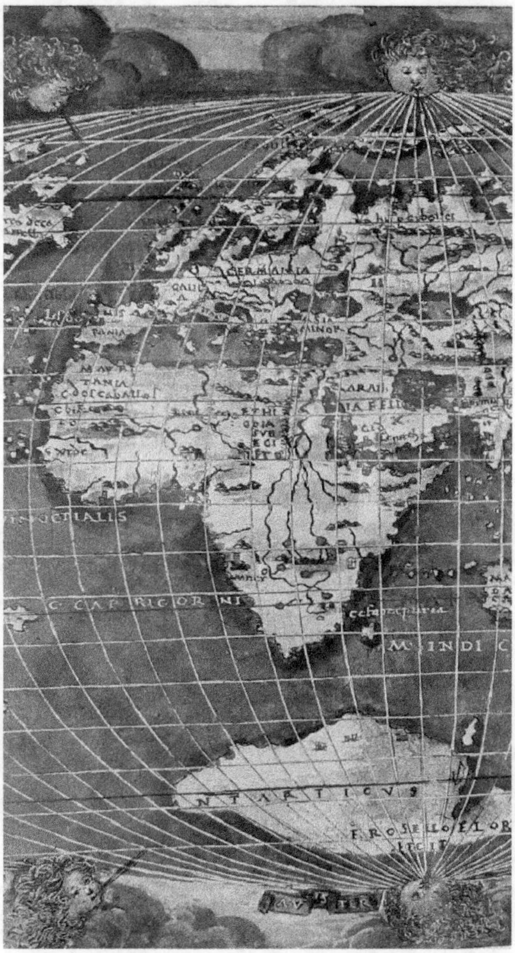

Contents

Introduction

Faster Than Light

In the annals of human history, few dreams have stirred the imagination as profoundly as the longing to transcend the boundaries of our world, to voyage beyond the stars, and to unravel the mysteries of time itself. For centuries, the notion of traveling faster than light has been relegated to the realm of science fiction, a tantalizing fantasy that seemed forever out of reach, bound by the unyielding constraints of Einstein's relativity. Yet today, we stand on the precipice of a new era, one where the impossible is becoming possible, where the barriers of light and time are no longer insurmountable. For the first time in human history, we can envision feasible faster-than-light travel, a prospect that promises to redefine our place in the cosmos and usher in an unprecedented Age of Exploration. This book, Faster Than Light, is an invitation to explore the boundless possibilities that are now within our grasp.

At the heart of this revolution lies a groundbreaking new theory of six-dimensional spacetime, published in April of this year by Dr. Gunther Kletetschka, a physicist whose work has already begun to reshape our understanding of the universe. Dr. Kletetschka's theory is nothing short of a paradigm shift, a dazzling intellectual achievement that unifies the seemingly irreconcilable domains of relativity and quantum field theory. For decades, physicists have wrestled with the challenge of reconciling these two pillars of modern physics, each governing its own realm—relativity describing the vast scales of stars and galaxies, quantum field theory ruling the subatomic world. Dr. Kletetschka's six-dimensional framework bridges this divide, offering a cohesive model of the universe that is as elegant as it is revolutionary. In doing so, he has not only advanced our theoretical understanding but also laid the foundation for technologies that were once thought to be the stuff of dreams.

Complementing this theoretical leap is the mathematics of the Alcubierre Drive, a concept first proposed by physicist Miguel Alcubierre in 1994. The Alcubierre Drive is a theoretical construct that envisions a spacecraft enveloped in a bubble of warped spacetime, contracting space ahead and expanding it behind, allowing the vessel to move faster than light without violating the principles of relativity. While Alcubierre's idea was initially dismissed as a mathematical curiosity, requiring exotic forms of matter and energy that seemed unattainable, Dr. Kletetschka's six-dimensional spacetime provides the missing piece. His theory offers a physical and mathematical framework that makes the Alcubierre Drive not merely plausible but potentially achievable. Together, these two pillars—Kletetschka's spacetime model and Alcubierre's warp mathematics—form the foundation upon which a working stardrive can be built.

The implications of this breakthrough are staggering. A functioning stardrive would allow humanity to traverse the vast distances between stars in mere days or weeks, rather than centuries or millennia. The galaxy, with its billions of stars and planets, would no longer be a distant frontier but a realm within our reach. The cost of developing such a technology will be considerable, requiring resources on the scale of a major government or a visionary corporation like SpaceX. Yet the investment is well within the realm of possibility. The technological and industrial capabilities of the modern world, combined with the relentless drive of human ingenuity, make the construction of a stardrive an achievable goal. What once seemed an extravagant dream is now a matter of engineering, funding, and determination.

Remarkably, the same physics and mathematics that enable faster-than-light travel also open the door to even more extraordinary possibilities: time travel and journeys to alternate timelines. Dr. Kletetschka's six-dimensional spacetime theory posits the existence of multiple timelines, parallel realities that branch from our own, each representing a different unfolding of events. This is not mere speculation but a natural consequence of the mathematical structure of his model, which allows for the manipulation of spacetime in ways that permit travel not only across vast distances but also through the fabric of time itself. The prospect of visiting the past, exploring potential futures, or even stepping into alternate versions of our universe is no longer confined to the pages of fiction. These possibilities, once the domain of philosophers and storytellers, are now grounded in rigorous science, poised to transform our understanding of reality.

Dr. Kletetschka's work marks him as a worthy successor to Albert Einstein, whose theories of relativity reshaped the 20th century and laid the groundwork for modern physics. Like Einstein, Kletetschka has dared to reimagine the fundamental nature of the universe, challenging long-held assumptions and offering a vision that is both bold and precise. His unification of relativity and quantum field theory is a milestone in human thought, a testament to the power of the human mind to probe the deepest mysteries of existence. Yet, unlike Einstein's theories, which described a universe we could only observe, Kletetschka's framework provides a blueprint for action, a guide to building technologies that can reshape our destiny.

The development of faster-than-light travel, enabled by this new understanding, promises to inaugurate a new Age of Exploration, one that dwarfs the voyages of Columbus, Magellan, or even the Apollo missions. The stars, long symbols of the unattainable, would become destinations. Humanity could establish outposts on distant worlds, encounter alien ecosystems, and perhaps even make contact with other intelligent beings. The cultural, scientific, and philosophical implications of such exploration are impossible to overstate. Just as the discovery of the New World transformed Europe, sparking centuries of innovation and exchange, the exploration of the galaxy would redefine what it means to be human. It would challenge our assumptions, broaden our horizons, and inspire generations to come.

Yet this new era will not come without challenges. The engineering hurdles of constructing a stardrive are formidable, requiring advances in materials science, energy production, and computational modeling. The ethical and societal implications of time travel and alternate timelines raise profound questions about responsibility, identity, and the nature of choice. How will we navigate the moral complexities of altering the past or exploring parallel realities? What will it mean to encounter versions of ourselves in other timelines? These are questions that demand careful consideration, even as we celebrate the technological triumphs that make them possible.

This book is a call to action. It is an invitation to dreamers, scientists, engineers, and visionaries to join in the effort to turn the impossible into reality. The path to faster-than-light travel will require collaboration across disciplines and industries. It will demand courage, creativity, and an unwavering commitment to pushing the boundaries of what is possible. But the rewards are beyond measure: a future where humanity is no longer confined to a single planet, a single time, or a single reality.

As we stand at this crossroads, the words of the poet Tennyson echo across the centuries: “To strive, to seek, to find, and not to yield.” The journey to the stars begins with the audacity to believe that we can go faster than light, that we can bend time, and that we can explore the infinite possibilities of the cosmos. Dr. Kletetschka’s theory and the Alcubierre Drive are the keys to this journey, tools that unlock a universe of wonder. This book is a guide to that journey, a map of the science, the vision, and the dreams that will carry us to the stars and beyond. The time has come to take the first step into a future that is faster than light.

Chapter I

A New Mathematical Framework for Reality

Picture time not as a single, unyielding line pulling you from birth to death, but as a vast, three-dimensional expanse, like an endless forest where you can wander forward, sideways, or even twist through unseen paths. This is the bold vision of Dr. Gunther Kletetschka, a geophysicist at the University of Alaska Fairbanks, whose 2025 paper in Reports on Advances in Physical Sciences proposes a universe built not on Einstein's four-dimensional spacetime—three dimensions of space plus one of time—but on a six-dimensional framework: three dimensions of time and three of space. Dr. Kletetschka argues that time, not space, is the fundamental canvas of existence, with space emerging as a secondary effect, like shadows cast by a flickering flame. This model does not merely tweak existing physics. It aims to unify the quantum world of particles with the cosmic realm of gravity, explain why particles have their precise masses, and offer testable predictions about mysteries like dark matter and the universe's expansion.

To appreciate Kletetschka's revolution, we need to start with the current standard picture of the universe, rooted in Albert Einstein's 1915 theory of general relativity. We experience three spatial dimensions daily: left-right, forward-back, and up-down. These form the stage where objects exist, move, and interact. Time, in contrast, feels like a one-way river, flowing relentlessly from past to future. Einstein wove these dimensions into a single four-dimensional fabric called spacetime, where massive objects like planets or stars bend the paths of other objects, creating the effect we call gravity. A star, for instance, warps spacetime like a bowling ball resting on a trampoline, causing planets to orbit along curved paths called geodesics. This model has been spectacularly successful for over a century, predicting everything from the precise orbit of Mercury to the bending of starlight during a solar eclipse. It even explains the behavior of black holes, where spacetime curves so intensely that light cannot escape.

However, this four-dimensional framework fails when we transition to the quantum scale of reality, where particles like electrons, quarks, and photons dance according to probabilities. The Standard Model of particle physics governs this realm, describing particles as excitations in invisible fields that permeate the universe. An electron, for example, is a ripple in the electron field, much like a wave on the ocean. Yet, the Standard Model leaves critical questions unanswered. Why do particles have the specific masses they do? The electron's mass is a precise value, measured in experiments, but the Standard Model offers no explanation for why it is precisely that number and not another. Similarly, the masses of quarks, which form protons and neutrons, seem arbitrarily set, requiring physicists to plug in 19 free parameters—numbers derived from experiments, not from first principles. Another puzzle is the hierarchy problem: why is gravity, which governs planets and galaxies, so much weaker than the electromagnetic or nuclear forces that dominate at the quantum level? A magnet can lift a paperclip against Earth's entire gravitational pull, highlighting this stark disparity.

Efforts to bridge quantum mechanics and gravity have led to ideas like string theory, which proposes extra spatial dimensions—often 10 or 11—curled up so tiny that we cannot detect them. These theories, while mathematically elegant, are notoriously hard to test, as the energies required to probe such dimensions are far beyond current technology. Kletetschka takes a different approach, suggesting that the problem lies not in adding more space but in rethinking time itself. His core idea is that time has three dimensions, not one, and that these three temporal dimensions, combined with the three spatial dimensions, form a six-dimensional universe that resolves many of physics' deepest mysteries.

What does three-dimensional time mean? Imagine the first time dimension as the familiar forward-backward flow we experience, like the timeline of your life from breakfast to bedtime. The second dimension is a sideways direction, representing alternative paths or possibilities—think of the different lives you might have led if you had chosen a different career or moved to another city. The third dimension acts as a kind of bridge or twist, allowing interactions between these branching paths without breaking the rule that causes must precede effects. Together with the familiar spatial dimensions—left-right, forward-back, and up-down—this creates a six-dimensional manifold, a smooth, flexible surface described by six coordinates. Why three time dimensions? Dr. Kletetschka explored configurations with one, two, or more time dimensions, ultimately concluding that three is the correct number. Fewer dimensions cannot account for the diversity of particles and forces we observe, while adding more risks introducing paradoxes, like effects happening before causes, which would unravel the observed logical structure of reality.

At human scales, these extra time dimensions are invisible, curled up into temporal loops so small—on the order of the Planck length, about 10^{-35} meters—that we perceive only the familiar flow of time. Picture a garden hose: from a distance, it looks like a one-dimensional line, but up close, it has a tiny circular cross-section. Similarly, the second and third time dimensions are compactified, noticeable only at the quantum scale or during extreme events like the Big Bang. At those scales, the full six-dimensional structure shapes the universe's behavior, from the masses of particles to the expansion of the cosmos.

The most radical part of Kletetschka's theory is that space is not the primary stage of reality—it is a consequence of time's geometry. In Einstein's relativity, gravity arises when mass-energy warps spacetime, like a heavy object depressing and stretching a rubber sheet. Kletetschka flips this script: the three-dimensional time manifold is the fundamental structure, and spatial dimensions emerge as folds or projections of temporal curvature. Imagine a crumpled piece of paper: the creases create patterns that appear as lines or surfaces from certain angles. In this view, concentrations of energy in the three-dimensional time manifold create ripples that we perceive as the spatial dimensions of left-right, forward-back, and up-down. This shift is profound because it suggests that quantum fields (which govern particles) and gravity (which governs spacetime) are two sides of the same coin, both arising from the dynamics of a temporal manifold.

This idea offers a path to unify quantum mechanics and gravity, a goal that has been the holy grail of modern physics for a hundred years. In the Standard Model, particles like electrons or quarks are described by quantum fields, but gravity remains an outsider, described by the smooth geometry of general relativity. Kletetschka's model bridges this gap by treating both as manifestations of temporal interactions. Quantum fields are vibrations in the three-dimensional time manifold, while gravity is the large-scale tension created by these vibrations. This unification could resolve the hierarchy problem, explaining why gravity is so weak. It is a diluted effect, filtered through the compactified second and third time dimensions, much like how a whisper carries less force than a shout.

One of Kletetschka's most compelling claims is that his model naturally predicts the masses of fundamental particles, a feat the Standard Model achieves only by manually inputting 19 arbitrary parameters. Consider the electron, whose mass is a precise value measured in particle accelerators. Kletetschka proposes that particle masses arise from waves oscillating in the three-dimensional time manifold, like musical notes played on a complex, multidimensional instrument. By modeling these waves with specific frequencies—drawn from experimental data—his framework calculates the exact masses of particles like the electron, the muon (a heavier cousin of the electron), and even quarks, which combine to form protons and neutrons. These masses are not random but emerge from the geometry of time, determined by how waves resonate in the extra temporal dimensions.

Electric charge also gets a new explanation. In standard physics, charge is a property particles possess, like a label saying "-1" for an electron or "+1" for a proton. Kletetschka sees charge as a topological feature—a kind of knot or twist in the temporal manifold that remains stable under deformation, like a loop in a rope that stays knotted no matter how you pull it. The discrete nature of charge (why it comes in exact units) arises because the geometry of the second and third time dimensions allows only certain stable configurations, locking charges into integer multiples. This topological approach could explain why particles have consistent charges across the universe, a mystery the Standard Model takes for granted.

Forces, too, are reimagined in this six-dimensional framework. Electromagnetism, which governs light, electricity, and magnetism, is described in standard physics by a set of equations developed by James Clerk Maxwell in the 19th century. Kletetschka extends these equations into six dimensions, introducing a new field tied to the third time dimension. This extra field predicts subtle changes in how particles interact, such as slight variations in the decay rates of particles like muons. These deviations are small but potentially measurable in high-energy experiments at facilities like CERN's Large Hadron Collider, expected to reach new levels of precision by 2030. Gravity, meanwhile, is no longer a curvature of spacetime but a tension in the three-dimensional time manifold. Massive objects create stress in this temporal fabric, which propagates as the spatial curvature we observe at large scales. This explains the weakness of gravity: it is a secondary effect, diluted by the compactified extra time dimensions, like a faint echo of a louder sound.

The theory's implications extend to the biggest questions in cosmology. According to current thinking, the universe is composed of roughly 5% ordinary matter (the stuff of stars and planets), 27% dark matter (invisible but gravitationally influential), and 68% dark energy (driving the universe's accelerating expansion). The nature of dark matter and dark energy remains one of physics' greatest puzzles. Kletetschka suggests that dark matter could be particles or fields residing in the second time dimension, interacting weakly with our familiar world but contributing to gravity's pull. Dark energy might arise from the third time dimension stretching out, pushing the spatial dimensions apart and driving cosmic expansion. The Big Bang, traditionally seen as a sudden explosion of space, is reimagined as a temporal event: a moment when the second and third time dimensions unfurled from their compactified state, giving rise to the spatial universe we observe. These ideas adjust the equations governing cosmic expansion, predicting values for the universe's growth rate and the cosmological constant (a number tied to dark energy) that align closely with observations, without the fine-tuning required in standard models.

What makes Kletetschka's theory stand out is its testability, a critical feature for any scientific idea. Unlike some speculative frameworks, like certain versions of string theory that require unreachable energies to test, Kletetschka's model makes concrete predictions. For example, it suggests that particles like muons or kaons (exotic particles produced in accelerators) will decay slightly differently due to interactions in the third time dimension. These differences could be detected in experiments at CERN or future colliders by the early 2030s. The theory also predicts a refined value for the cosmological constant, which could be tested by studying the cosmic microwave background—the faint radiation left over from the Big Bang—using telescopes like the Simons Observatory or the upcoming CMB-S4 experiment. Even gravitational waves, ripples in spacetime detected by observatories like LIGO, might show additional patterns or polarizations due to the extra time dimensions, a signal that future detectors like the Einstein Telescope could pick up.

Predictably, Dr. Kletetschka's work has raised concerns in the scientific community. His academic background as a geophysicist, not a theoretical physicist, makes some in the physics community wary. The field of quantum gravity is dominated by specialists like those working on string theory or loop quantum gravity, and an outsider's perspective can face resistance. His idea of extra time dimensions is not entirely new; it echoes earlier proposals, like physicist Itzhak Bars' two-time physics from the 1990s, which struggled with issues like causality (ensuring effects follow causes). Kletetschka's model avoids these pitfalls by carefully designing the geometry of the extra time dimensions to preserve positive energy and causality, but the compactified nature of these dimensions makes them hard to detect directly. Current experiments lack the precision to probe effects at the Planck scale, though technological advances expected in the next decade could change this. Critics also point out that the theory needs more development to fully replace the Standard Model or general relativity, as it must reproduce all their successful predictions while adding new insights.

Despite these challenges, Kletetschka's work is grounded in empirical rigor. His predictions about particle masses align with experimental data, and his cosmological adjustments match observations without the arbitrary tweaks often needed in standard models. The theory's elegance lies in its simplicity: by rethinking time, it offers a unified explanation for phenomena that span the full gamut of reality, from the subatomic to the cosmic. It also sparks profound philosophical questions. If time is three-dimensional, what does that mean for free will or destiny? The second time dimension, with its branching possibilities, suggests a universe where multiple outcomes coexist, challenging our linear sense of history. The third dimension, acting as a bridge, raises the possibility of interactions across these branches, perhaps influencing events in ways we cannot yet imagine. Could we one day detect these extra dimensions or even manipulate them? Such ideas evoke science fiction, but they are rooted in the same bold thinking that led to relativity or quantum mechanics—ideas once dismissed as outlandish but now foundational to physics.

The technological implications are equally tantalizing. If Kletetschka's model is correct, manipulating the extra time dimensions could open new frontiers. Consider propulsion systems that exploit temporal geometry to bypass spatial distances, akin to wormholes but grounded in time's structure. Such an approach could make feasible interstellar travel via models such as the Alcubierre Drive. Even time travel might be possible. We will explore these possibilities in subsequent chapters.

For the non-specialist, Kletetschka's theory is a reminder that science thrives on questioning the obvious. Time, which we measure with clocks and calendars, may be far more complex—a multidimensional tapestry weaving together particles, forces, and the cosmos itself. As experiments advance, we may soon learn whether we are trapped in a single timeline or potential explorers of a much richer temporal landscape. Kletetschka's vision challenges us to see time not as a fleeting arrow but as the very fabric of existence, inviting us to wonder about what other secrets the universe may hold in its hidden dimensions.

Mathematical Appendix

The following equations, referenced in the essay, provide the mathematical foundation of Kletetschka's six-dimensional spacetime theory, explained clearly for non-specialists.

1. **Minkowski Metric (Standard 4D Spacetime)**

$$ds^2 = c^2dt^2 - dx^2 - dy^2 - dz^2$$

This measures the "distance" between two nearby events in Einstein's four-dimensional spacetime. *ds* is the infinitesimal spacetime interval, *c* is the speed of light (approximately 299,792,458 meters per second), and *dt*, *dx*, *dy*, *dz* are tiny changes in time and the three spatial coordinates. The positive sign for time and negative signs for space ensure time's unique role, preserving causality through light cones.

1. **Kletetschka's Six-Dimensional Metric**

$$ds^2 = dt_1^2 + dt_2^2 + dt_3^2 - dx^2 - dy^2 - dz^2$$

This extends the Minkowski metric to six dimensions, with three time dimensions (t_1, t_2, t_3) and three spatial dimensions (x, y, z). Positive signs for time and negative for space maintain positive energy and causality, allowing four-dimensional spacetime to emerge when t_2 and t_3 are compactified.

1. **Six-Dimensional Einstein Field Equation**

$$R_{\mu\nu} - \frac{1}{2}g_{\mu\nu}R = \frac{8\pi G}{c^4}T_{\mu\nu}$$

This generalizes Einstein's gravity equation to six dimensions. *R_{|mu|nu}* is the Ricci curvature tensor, describing the manifold's curvature; *g_{|mu|nu}* is the metric tensor encoding geometry; *R* is the scalar curvature; *G* is the gravitational constant; *c* is the speed of light; and *T_{|mu|nu}* is the stress-energy tensor, representing mass and energy. It models how temporal curvature generates space and forces.

1. Temporal Wave Equation for Particle Masses

$$\nabla^2\psi - \frac{1}{c^2}\frac{\partial^2\psi}{\partial t_1^2} + \frac{\partial^2\psi}{\partial t_2^2} + \frac{\partial^2\psi}{\partial t_3^2} = -\frac{m^2c^2}{\hbar^2}\psi$$

This describes particles as waves in the three-dimensional time manifold. ψ is the wavefunction (quantum probability field); ∇^2 is the Laplacian for spatial curvature; $\partial^2/\partial t_1^2$, $\partial^2/\partial t_2^2$, $\partial^2/\partial t_3^2$ describe temporal oscillations; m is the particle's mass; c is the speed of light; and $\hbar$ is the reduced Planck constant (6.626 × 10^{-34} J·s / 2π). Solving this with compactified t_2 and t_3 yields particle masses.

---

1. Electron Mass Formula

$$m_e = \frac{\hbar\omega}{c^2}\sin\left(\frac{\pi t_2}{\tau_1}\right)$$

This predicts the electron's mass ($m_e = 0.511$ MeV/c^2). $\hbar$ is the reduced Planck constant, ω is the angular frequency, c is the speed of light, and the $\sin$ term modulates t_2 relative to t_1, matching experimental data.

---

1. Modified Friedmann Equation

$$\left(\frac{\dot{a}}{a}\right)^2 = \frac{8\pi G}{3}\rho + \frac{\Lambda}{3} + \frac{k}{a^2}$$

This governs cosmic expansion. a is the scale factor (universe's size), ρ is energy density, Λ is the cosmological constant (tied to dark energy), G is the gravitational constant, and k reflects spatial curvature. The t_3 contribution refines Λ to match observations.

Chapter II

Feasible Faster Than Light Travel

Imagine a universe where faster-than-light (FTL) travel, a concept often explored in speculative discussions and science fiction, is theoretically possible and practically feasible, thanks to a radically new understanding of time itself. This provocative vision has become a reality thanks to Dr. Gunther Kletetschka, a geophysicist at the University of Alaska Fairbanks, who appears to have succeeded in surpassing Einstein's Theory of General Relativity.

In an April 2025 paper published in Reports in Advances of Physical Sciences, Kletetschka proposed a six-dimensional spacetime framework consisting of three time dimensions and three spatial dimensions, with time as the primary structure and space as a derived effect, like ripples on a pond. His theory permits us to take a fresh look at the Alcubierre Drive, a theoretical FTL concept introduced by physicist Miguel Alcubierre in 1994. This essay explores how Kletetschka's model allows us to reimagine the Alcubierre Drive, particularly its reliance on negative energy density—heretofore a critical theoretical hurdle in the way of faster-than-light travel.

To understand why faster-than-light travel now appears attainable, let us first examine the Alcubierre Drive within Einstein's Theory of General Relativity. In our daily reality, we experience three spatial dimensions—left-right, forward-back, and up-down—forming the stage where objects exist and move. Time feels like a one-way flow, carrying us from past to future. General Relativity weaves these dimensions into a four-dimensional spacetime, in which massive objects like stars bend the fabric of reality, creating gravity. This framework is based on a cosmic speed limit: nothing can travel faster than light, which moves at about 300 million meters per second. Alcubierre's 1994 proposal sidesteps this limit by manipulating spacetime itself. His "warp drive" creates a localized spacetime bubble around a spaceship, in which spacetime inside remains flat, and the ship stays still relative to its local surroundings. Outside, spacetime contracts in front of the bubble and expands behind, propelling the ship faster than light relative to the universe. Much like a surfer standing motionless on a surfboard while riding a rapidly moving wave toward the shore, the ship remains motionless within its bubble, while the bubble itself "surfs" the fabric of spacetime at incredible speeds far exceeding those of light.

This concept depends on a specific spacetime geometry that requires negative energy density. Energy density measures how much energy is packed into a volume. Ordinary matter, like stars or people, has positive energy density, contributing to gravitational attraction, pulling things together. Negative energy density, however, pushes spacetime outward, creating a repulsive effect essential for the Alcubierre Drive. This repulsion contracts space ahead of the bubble and expands it behind, enabling faster-than-light motion. In standard physics, negative energy density is exotic and rare. Quantum mechanics suggests it is possible in tiny amounts, such as in the Casimir effect, where vacuum fluctuations between uncharged metal plates create a small negative energy region due to restricted quantum waves. However, the Alcubierre Drive demands vast quantities—potentially equivalent to the mass of Jupiter compressed into a small volume—making it impractical. Some physicists argue that negative energy violates principles like the weak energy condition, which suggests energy density should be non-negative for all observers, though quantum effects allow limited exceptions. Stability is another hurdle, as negative energy regions can collapse or dissipate, disrupting the warp bubble.

Kletetschka's six-dimensional spacetime model offers a potential solution by redefining how energy and spacetime interact. Instead of four dimensions, he proposes six: three time dimensions and three spatial ones. The first time dimension is the familiar forward-backward flow we experience, like the progression from breakfast to bedtime. The second represents a sideways direction, akin to branching possibilities or alternate timelines—think of the different lives you might have led by choosing a different career or city. The third acts as a bridge, facilitating transitions between these branches while ensuring causality, so effects always follow causes. At human scales, the second and third time dimensions are curled up tightly, on the order of the Planck length—a billionth of a trillionth of a trillionth of a meter—making them imperceptible, like the cross-section of a thin wire seen from afar. At quantum scales, such as inside atoms, or during cosmic events like the Big Bang, these dimensions become significant, shaping the universe's fundamental behavior.

The core innovation is that space emerges from the curvature of the three-dimensional time manifold. Imagine a crumpled piece of paper: the creases form patterns that appear as lines or surfaces from certain angles. Similarly, energy concentrations in the temporal manifold create ripples that we perceive as the spatial dimensions of left-right, forward-back, and up-down. This shift is transformative for the Alcubierre Drive because it suggests that manipulating the temporal manifold could directly produce the spacetime geometry needed for the warp bubble, potentially bypassing the need for exotic negative energy. Instead of bending four-dimensional spacetime with vast amounts of negative energy, Kletetschka's model allows us to sculpt the temporal manifold, letting spatial effects emerge naturally.

The negative energy density required for the Alcubierre Drive is a geometric challenge in standard physics. Kletetschka's model addresses this through its temporal wave function, which describes how energy fluctuates across the six-dimensional manifold. The wave function is a quantum field, a mathematical description of the probability of finding energy or particles at specific points in the six-dimensional space. It may be compared to a map showing where energy is likely to ripple in a vast, multidimensional ocean. The wave function includes a mathematical term called the Laplacian, which measures how the wave function curves or bends across the three spatial dimensions. Think of the Laplacian as a tool that checks how much a surface fluctuates at a given point, capturing the shape of the wave in space.

The wave function also includes second partial derivatives with respect to each time dimension. A second partial derivative measures the rate of change of the rate of change of the wave function. For the first time dimension, this is like measuring how quickly the wave's slope changes as time progresses—similar to the acceleration of a vibrating string. The second and third time dimensions have similar terms, describing how the wave oscillates across branching possibilities and transitions between them. These oscillations create complex patterns, like waves rippling through a three-dimensional pond, with the second dimension allowing energy to fluctuate across alternate timelines and the third stabilizing these fluctuations to maintain the coherence of the structure as a whole.

In the context of the Alcubierre Drive, these temporal oscillations are critical. The second time dimension's branching structure allows energy to flow across different possible timelines, creating a "temporal flux" that, when projected into the spatial dimensions, mimics the negative energy density needed for the warp bubble. This is not true negative energy but a geometric effect, like a shadow creating a shape not present in the original source object. The third time dimension, acting as a mediator, ensures these fluctuations remain stable, preventing the warp bubble from collapsing—a common problem in the standard Alcubierre model where negative energy regions are inherently unstable. This stability could significantly reduce the energy requirements of the Alcubierre Drive, from the equivalent of Jupiter's mass to something potentially achievable with advanced technologies. We will discuss this aspect of the drive in subsequent chapters.

To understand the wave function's role more deeply, consider its physical interpretation. The wave function describes energy as a wave propagating through the six-dimensional manifold, with each time dimension contributing a unique dynamic. The first time dimension governs the familiar progression of events, ensuring the wave evolves in a way that aligns with our experience of time. The second dimension introduces a multiplicity of paths, allowing the wave to explore different configurations, like a story with multiple endings. The third dimension ties these paths together, acting like a conductor orchestrating a symphony, ensuring that the wave's fluctuations do not lead to chaotic or paradoxical outcomes. The second derivatives in the wave function capture the curvature of these temporal waves, reflecting how their behavior accelerates or decelerates in each dimension. This curvature is what allows the temporal manifold to produce the complex geometries needed for the

Alcubierre Drive, effectively creating regions where space contracts and expands without requiring exotic matter.

Practically, creating the warp bubble in Kletetschka's framework involves exciting the temporal manifold to produce these wave patterns. His mathematical model extends electromagnetism, as defined by James Clerk Maxwell's equations, to include a scalar field influenced by the third time dimension, with a first partial derivative capturing how this field changes along the bridging dimension. This means that advanced devices, such as powerful pulsed or oscillating magnets, could generate high-frequency oscillations in the second and third time dimensions, inducing the temporal curvature needed for the Alcubierre Drive's geometry. Unlike the Casimir effect, which produces tiny negative energy regions through quantum fluctuations, this approach leverages the temporal manifold's dynamics, potentially requiring less energy and offering a more practical path to faster-than-light travel. Such a device could create controlled temporal waves, tuned to resonate in the second and third time dimensions, producing the exact spacetime curvature needed for the warp bubble.

Cosmologically, Kletetschka's model connects the Alcubierre Drive to the universe's expansion, which is driven by dark energy—a mysterious force causing the cosmos to accelerate apart. The third time dimension's stretching mirrors the warp bubble's dynamics, suggesting that faster-than-light travel could harness mechanisms similar to those shaping the universe. During the Big Bang, the universe underwent rapid inflation, a period of exponential expansion driven by a field similar to dark energy. Kletetschka's model posits that this expansion arose from the unfurling of the second and third time dimensions, creating the spatial universe we observe. By tapping into these temporal mechanisms, the Alcubierre Drive could create localized expansions and contractions of space, mimicking the early universe's dynamics on a smaller scale. This connection not only makes FTL travel theoretically plausible but also ties it to the fundamental processes governing cosmic evolution. Just as a sailboat is propelled by the winds and currents that shape the oceans and seas, so the Alcubierre Drive could be powered by the fundamental geometry and forces of the universe.

This expanded conceptual framework also offers insights into how the three time dimensions interact to produce these effects. The first time dimension provides the baseline progression, anchoring the wave function's evolution in a way that aligns with our everyday experience. The second dimension's branching structure introduces a richness to the wave function, allowing it to explore multiple configurations simultaneously, like a quantum particle existing in multiple states. This multiplicity is crucial for generating the complex energy patterns needed to mimic negative energy density. The third dimension, as a mediator, ensures that these configurations remain coherent, preventing the wave function from producing paradoxical outcomes, such as effects preceding causes. Together, these dimensions create a dynamic interplay that allows the temporal manifold to sculpt spacetime in ways that standard four-dimensional models cannot support.

As noted in Chapter One, Kletetschka's model is testable, which is a significant advantage. It predicts subtle deviations in particle behaviors, such as how quickly particles like muons decay, influenced by the third time dimension's scalar field. These deviations could be detected at high-energy facilities like CERN's Large Hadron Collider, which is expected to achieve greater precision by the early 2030s. The model also suggests refinements to the cosmological constant—the number tied to dark energy—which could be verified by studying the cosmic microwave background, the faint radiation from the Big Bang, using telescopes like the Simons Observatory or the upcoming CMB-S4 experiment. Gravitational waves, ripples in spacetime detected by observatories like LIGO, might show additional patterns due to the extra time dimensions, a signal that future detectors like the Einstein Telescope could pick up. These tests are crucial for validating the model and its implications for interstellar travel.

Significant challenges remain, however. Accessing the compactified second and third time dimensions requires probing effects at the Planck scale, a realm far beyond current technology. While advancements expected by the 2030s could bridge this gap, they demand breakthroughs in quantum gravity or high-energy physics. The Alcubierre Drive, even in Kletetschka's framework, may risk creating paths that could allow time paradoxes, like traveling back to prevent an event. The third time dimension's role in preserving causality mitigates this, but the exact mechanisms need further exploration. Energy demands, while potentially reduced by the temporal approach, may still be substantial, requiring innovative engineering to generate and sustain the necessary temporal oscillations.

Philosophically, the three-dimensional time manifold challenges our understanding of reality. The second time dimension, with its branching possibilities, suggests a universe where multiple outcomes coexist, raising questions about whether a faster-than-light ship might be able to navigate these branches, selecting a different cosmic history. Imagine a warp bubble not just crossing space but surfing through alternate timelines, choosing a path where a distant star system is reachable in moments. The third time dimension, as a mediator, might allow controlled interactions across these branches, offering a new perspective on free will and destiny. Could we influence events across timelines, or are we bound to a single path? These questions revive the ancient link between science and philosophy, inviting us to rethink our place in a multidimensional cosmos.

Technologically, the implications are transformative. If Kletetschka's model is correct, faster-than-light travel could enable interstellar journeys in days or weeks instead of centuries or millennia, making the colonization of distant planets or contact with extraterrestrial civilizations feasible. Beyond travel, manipulating the temporal manifold could revolutionize technology. Energy storage in temporal pockets—regions of the second or third time dimensions—could lead to batteries far more efficient than today's. Quantum computing might leverage the branching nature of the second time dimension for parallel processing across multiple timelines, exponentially increasing computational power. Communication systems could use temporal waves to send signals through the third dimension, bypassing spatial limitations. These possibilities, while speculative, highlight the potential for Kletetschka's theory to inspire a new era of innovation, much as Einstein's relativity led to a world of new technologies in the 20th century.

The expanded view of Kletetschka's wave function also underscores its broader significance. In standard physics, wave functions describe particles or fields in four-dimensional spacetime, but Kletetschka's version operates in a six-dimensional manifold, incorporating the unique dynamics of three time dimensions. The second derivatives in each time dimension allow the wave function to capture complex temporal behaviors, such as oscillations that resonate across alternate timelines or stabilize through transitions. This richness enables the model to explain not only the Alcubierre Drive's geometry but also fundamental phenomena like particle masses and forces. For example, the wave function's ability to produce discrete energy states in the temporal manifold could explain why particles like electrons have specific masses, a mystery in standard physics. By extending this to the Alcubierre Drive, the wave function becomes a versatile tool for sculpting spacetime.

The idea of traveling faster than light, once confined to speculative fiction, could well become a reality thanks to reimagining time as a three-dimensional landscape. By addressing the negative energy problem through temporal geometry, Kletetschka's model brings us closer to a universe where interstellar voyages are not just dreams but a feasible engineering challenge. As experiments test his predictions, we may find that the key to crossing the stars lies not in conquering space but in mastering the hidden depths of time's multidimensional geometry.

Mathematical Appendix

The following equations underpin **Kletetschka's** six-dimensional spacetime theory and its implications for the Alcubierre Drive, explained in plain English.

Minkowski Metric (Standard 4D Spacetime)

$$ds^2 = c^2dt^2 - dx^2 - dy^2 - dz^2$$

This measures the "distance" between two nearby events in Einstein's four-dimensional spacetime. ds is the infinitesimal spacetime interval, c is the speed of light (approximately 299,792,458 meters per second), and dt, dx, dy, dz are tiny changes in time and the three spatial coordinates. The positive sign for time and negative signs for space ensure causality via light cones.

Kletetschka's Six-Dimensional Metric

$$ds^2 = dt_1^2 + dt_2^2 + dt_3^2 - dx^2 - dy^2 - dz^2$$

This defines the six-dimensional universe with three time dimensions (t_1, t_2, t_3) and three spatial dimensions (x, y, z). Positive signs for time and negative for space maintain positive energy and causality, allowing four-dimensional spacetime to emerge when t_2 and t_3 are compactified.

Six-Dimensional Einstein Field Equation

$$R_{\mu\nu} - \frac{1}{2} g_{\mu\nu} R = \frac{8\pi G}{c^4} T_{\mu\nu}$$

This generalizes Einstein's gravity equation to six dimensions. $R_{\mu\nu}$ is the Ricci curvature tensor, describing manifold curvature; $g_{\mu\nu}$ is the metric tensor; R is the scalar curvature; G is the gravitational constant; c is the speed of light; and $T_{\mu\nu}$ is the stress-energy tensor. It models how temporal curvature generates the Alcubierre geometry.

Alcubierre's Warp Drive Metric

$$ds^2 = c^2 dt^2 - [dx - v_s(t) f(r_s) dt]^2 - dy^2 - dz^2$$

This describes the warp bubble's geometry. $v_s(t)$ is the bubble's velocity, $f(r_s)$ is the shape function (1 inside, 0 outside), and r_s is the distance from the ship. It requires negative energy density to function.

Stress-Energy Tensor for Negative Energy Density

$$T_{00} = -\frac{c^4}{8\pi G}\left(\frac{v_s^2\rho^2}{4r_s^2}\right)\left(\frac{df}{dr_s}\right)^2$$

This shows the negative energy density ($T_{00} < 0$) needed for the Alcubierre bubble, where ρ is the radial coordinate. Kleteetschka's temporal curvature mimics this effect geometrically.

Temporal Wave Equation

$$\nabla^2\psi - \frac{1}{c^2}\frac{\partial^2\psi}{\partial t_1^2} + \frac{\partial^2\psi}{\partial t_2^2} + \frac{\partial^2\psi}{\partial t_3^2} = 0$$

This describes energy fluctuations in the temporal manifold. ψ is the wavefunction; ∇^2 is the Laplacian, defined as $\nabla^2\psi = \partial^2\psi/\partial x^2 + \partial^2\psi/\partial y^2 + \partial^2\psi/\partial z^2$, measuring spatial curvature; $\partial^2\psi/\partial t_1^2$, $\partial^2\psi/\partial t_2^2$, $\partial^2\psi/\partial t_3^2$ are second partial derivatives, capturing the rate of change of the wave's slope in each time dimension; c is the speed of light; and the zero right-hand side indicates a free wave. Oscillations in t_2 and t_3 mimic negative energy density.

Modified Wave Equation for Particle Masses

$$\nabla^2\psi - \frac{1}{c^2}\frac{\partial^2\psi}{\partial t_1^2} + \frac{\partial^2\psi}{\partial t_2^2} + \frac{\partial^2\psi}{\partial t_3^2} = \frac{m^2c^2}{\hbar^2}\psi$$

This version relates to particle masses, where m is the mass and $\hbar$ is the reduced Planck constant (6.626×10^{-34} J·s$/2\pi$). It shows how temporal waves generate particle properties.

Chapter III

The Stardrive

Beneath the starry expanse of the cosmos, humanity stands poised to transcend the boundaries of space and time through faster-than-light (FTL) travel, enabled by Dr. Gunther Kletetschka's six-dimensional theory of spacetime and a reimagined version of the Alcubierre Drive. At the heart of a future starship's ability to achieve FTL travel is Kletetschka's temporal wave function, which allows high-frequency laser light pulses to excite the t2 (branching possibilities) and t3 (transitions) axes of three-dimensional time, creating the spacetime curvature required for the Alcubierre Drive's warp bubble. This essay explores how the temporal wave function underpins the operation of a hypothetical stardrive, the Temporal Wave Generator (TWG), focusing on the critical role of laser light pulses and the wave function in creating a 50-meter warp bubble.

The Alcubierre Drive and Kletetschka's Six-Dimensional Framework

The Alcubierre Drive, proposed by Miguel Alcubierre in 1994, envisions a spacecraft encased in a warp bubble that contracts spacetime ahead and expands it behind, enabling effective FTL travel without violating local relativity. In standard physics, this requires negative energy density—a state where energy is less than zero, unlike ordinary matter, which has positive energy (e.g., a 1-kilogram object with a rest energy of about 90,000 trillion joules). Exotic matter with negative energy density or quantum effects, like the Casimir effect (where vacuum fluctuations between plates create slight negative pressure), were once thought necessary. Kletetschka's 2025 theory, published in Reports in Advances of Physical Sciences, offers an alternative by leveraging a six-dimensional spacetime model with three spatial dimensions (x, y, z) and three time dimensions: t1 (linear time), t2 (branching possibilities, like quantum superpositions), and t3 (transitions between states). In this model, time is the primary framework, and spatial dimensions emerge as projections of temporal curvature, enabling spacetime distortions like the warp bubble without exotic matter.

Kletetschka's theory introduces a temporal wave function that describes energy fluctuations across the six-dimensional manifold, governing how temporal curvature projects spatial effects. The TWG, an 18-meter-diameter, 2.5-meter-thick toroidal ring on Deck 4 of the starship, implements this by using ytterbium-doped lasers to produce 1.5-petahertz (1.5 quadrillion cycles per second) light pulses, enhanced by quantum dots embedded in eight niobium-titanium superconducting coils generating 15-Tesla magnetic fields at 3.5 Kelvin. These pulses resonate with t2 and t3, creating a 50-meter warp bubble that envelops the ship, enabling travel at 50 times the speed of light (50c), making distant stars like Proxima Centauri b reachable in weeks or months.

The Role of the Temporal Wave Function in the TWG: A Conceptual Foundation

Kletetschka's theory posits that the universe's fundamental structure is a three-dimensional time manifold, with spatial dimensions emerging as projections of temporal curvature. The temporal wave function, a quantum field, describes energy distributions across t1, t2, and t3, with its Laplacian (summing second partial derivatives in spatial dimensions) and temporal derivatives capturing curvature dynamics. This wave function enables the TWG to manipulate the t2 and t3 dimensions, producing spacetime distortions for the warp bubble.

Imagine the laser pulses as a high-pitched hum, vibrating so rapidly that they resonate with the fabric of time itself. Kletetschka's wave function acts like a mathematical map, guiding the laser pulses to excite t2 and t3, bending time in a way that reshapes space around the starship. This eliminates the need for negative energy density, as the temporal curvature mimics the same effect, creating a bubble that shrinks space ahead and stretches it behind.

The TWG's design is optimized for this process. Its 18-meter-diameter, 2.5-meter-thick toroidal shape, chosen for its symmetry, ensures an even spacetime distortion, minimizing energy waste and structural stress while enveloping the entire starship in a 50-meter warp bubble. The torus houses eight segments, each with a niobium-titanium superconducting coil generating a 15-Tesla magnetic field, cooled to 3.5 Kelvin (-269.65°C) by a 200-kilowatt liquid helium cryocooler system. Niobium-titanium was selected for its ability to maintain superconductivity at achievable cryogenic temperatures and its high critical magnetic field, allowing stable, powerful magnetic fields. These fields support the laser pulses by creating a controlled environment, while quantum dots embedded in the coils fine-tune the pulses to achieve the precision needed for temporal resonance.

How the TWG Uses Light Pulses to Create the Warp Bubble

In the context of the starship, the TWG appears as a futuristic ring 18 meters wide with a metallic sheen reflecting the ship's high-tech interior. Its eight niobium-titanium coils are embedded with quantum dots—tiny, 4-to-6-nanometer particles of indium arsenide (for t2), gallium arsenide (for t3), and cadmium selenide (for t1)—acting like miniature tuning devices. Ytterbium-doped lasers, mounted along the torus's inner rim or a central hub, fire bursts of light pulsing 1.5 quadrillion times per second, with a power of 1.2 quadrillion watts, enough to briefly power a small city.

The temporal wave function enables the laser pulses to interact with the t2 and t3 dimensions. In Kletetschka's model, the wave function's oscillations in t2 (branching possibilities) and t3 (transitions) create temporal ripples that project spatial distortions, forming the 50-meter warp bubble that envelops the starship, contracting space ahead and expanding it behind. The wave function's second partial derivatives in t2 and t3 guide the pulses to resonate with these dimensions, like a key turning a lock.

The quantum dots are essential for precision. These particles, smaller than a virus, trap electrons in specific energy levels, like tiny cages, allowing them to absorb and re-emit light at exact wavelengths. Each type of dot is tuned to a time dimension: indium arsenide for t2's branching possibilities, gallium arsenide for t3's transitions, and cadmium selenide for t1's linear flow. When the laser pulses hit the dots, they sharpen the light to 99.999% coherence, ensuring the waves are perfectly aligned, like a laser pointer compared to a flickering bulb. This precision allows the pulses to hit the exact "frequency" needed to excite t2 and t3, maximizing the warp bubble's stability and efficiency.

The 15-Tesla magnetic fields support this process by creating a stable, high-energy environment. In standard physics, magnetic fields can influence charged particles and electromagnetic waves. In the TWG, the magnetic fields stabilize the quantum dots and focus the laser pulses, ensuring their interaction with the temporal dimensions is effective. The toroidal shape ensures these fields are distributed evenly, minimizing energy waste and structural stress, while the 50-meter warp bubble fully encloses the starship, allowing seamless FTL travel.

Enhancing the Warp Bubble with Quantum Systems

To boost the warp bubble's effect, the TWG includes a Bose-Einstein condensate (BEC) of 2 million rubidium atoms cooled to 8 nanokelvin, housed in a 6-meter octahedral chamber within a Penning trap using a 30-Tesla magnetic field and an ultra-high vacuum (a millionth of a billionth of Earth's atmospheric pressure). The BEC, where atoms merge into a single quantum wave, acts as a "temporal lens," amplifying the laser pulses' effects on the t2 dimension, enhancing branching configurations that strengthen the bubble. Kletetschka's wave function describes how this system evolves in six-dimensional spacetime, ensuring the BEC aligns the pulses' temporal resonance.

Powering the TWG

The TWG's immense energy needs are met by an antimatter-catalyzed fusion reactor, using 0.6 grams of antiprotons stored in a Penning trap (25 Tesla, ultra-high vacuum) to fuse 60 kilograms of deuterium and helium-3, a rare isotope mined from lunar soil. Each burst releases 50,000 trillion joules, equivalent to 12 megatons of TNT, powering the lasers and magnets. Heat is dissipated by 500 square meters of graphene radiators, capable of shedding 2,800 watts per square meter per degree of temperature difference. Backup power comes from a 6-megawatt perovskite solar array (300 square meters, 32% efficient) and a 600-kilowatt generator using 15 kilograms of plutonium-238, a radioactive material that generates heat for electricity.

Operational Dynamics and Control

The TWG creates a 50-meter warp bubble, enabling travel at 50c (15 billion meters per second). The starship's rotating decks simulate 0.7g gravity, ensuring crew comfort. Curva, a 600-qubit artificial intelligence, navigates using advanced mathematics, exploring millions of spacetime paths simultaneously. The pulses are stabilized by 25 laser interferometers (measuring distances to a billionth of a billionth of a meter) and 10 gravimeters (detecting tiny gravity changes). A timeline detector, sensitive to a trillionth of a trillionth of t1, prevents time-travel paradoxes, keeping the starship on a safe course.

This system enables remarkable journeys. A trip to Proxima Centauri b, some 40 trillion kilometers from Earth, takes just 31 days from the perspective of the crew and Earth observers, compared to 73,000 years by conventional rocket. The TWG's 18-meter toroidal design, with its eight niobium-titanium coils, ensures a uniform 50-meter bubble, balancing efficiency and structural integrity for such voyages.

Implications for Interstellar Exploration

Kletetschka's temporal wave function is the cornerstone of the TWG, enabling laser light pulses to bend the t2 and t3 dimensions, creating a warp bubble without exotic matter. The quantum dots and magnetic fields enhance this process, but the wave function allows the pulses to manipulate time, projecting spatial distortions. This opens the cosmos to humanity, with over 2000 potentially habitable planets within 100 light-years, including Proxima Centauri b (4.24 light-years) and TRAPPIST-1's seven planets (40 light-years). The TWG's light-based approach, supported by its compact, symmetrical design, makes FTL travel feasible, opening the way for a galactic civilization.

Mathematical Annex

Alcubierre Metric

This describes the spacetime geometry of the warp bubble:

$$ds^2 = -dt_1^2 + [dx - v_s(t)f(r_s)dt_1]^2 + dy^2 + dz^2$$

- $v_s(t)$: Bubble velocity ($50c = 1.5 \times 10^{10}$ m/s).
- $f(r_s)$: Shape function (1 inside the bubble, 0 outside).
- r_s: Radial distance (meters).

Kletetschka's Temporal Wave Equation (Free Waves)

This governs the TWG's quantum wavefunction:

$$\nabla^2\psi - \frac{1}{c^2}\frac{\partial^2\psi}{\partial t_1^2} - \frac{\partial^2\psi}{\partial t_2^2} - \frac{\partial^2\psi}{\partial t_3^2} = 0$$

- ψ: Quantum wavefunction.
- $\nabla^2\psi$: Spatial Laplacian (1/m²).
- $\partial^2\psi/\partial t_1^2, \partial^2\psi/\partial t_2^2, \partial^2\psi/\partial t_3^2$: Temporal accelerations (1/s²).

Modified Temporal Wave Equation (Particles)

This includes a mass term for the starship:

$$\nabla^2\psi - \frac{1}{c^2}\frac{\partial^2\psi}{\partial t_1^2} - \frac{\partial^2\psi}{\partial t_2^2} - \frac{\partial^2\psi}{\partial t_3^2} = \frac{m^2c^2}{\hbar^2}\psi$$

- m: Starship mass (~ 10^8 kg).
- $m^2c^2/\hbar^2$: Mass term (~ 10^68 kg²/m²).

Chapter IV

Building the First Starship: The Argo

In the quiet expanse of the cosmos, where stars burn with ancient fire and the void whispers of secrets untold, humanity stands on the precipice of a grand adventure. The Argo, a 50-meter prototype starship envisioned to carry a crew of ten to twenty to the distant surface of Proxima Centauri b, embodies this audacious dream. Drawing on Dr. Gunther Kletetschka's revolutionary six-dimensional spacetime model, the Argo promises to traverse 4.24 light-years in just one to five months, propelled by an Alcubierre-like warp bubble that bends the fabric of reality itself. The Argo is a pragmatic marvel, estimated to cost between $14 and $52 billion and requiring 15 to 20 years to build. This essay outlines how a starship like the Argo could be built, detailing the costs, timeline, and technological leaps needed to forge this vessel of exploration. It is a tale of human ingenuity, where science and aspiration converge to sail the cosmic seas.

The Argo's design is a study in balance, a vessel modest in scale yet boundless in ambition. Measuring just 50 meters in length and 20 meters in width, with a mass of 2,000 to 3,000 metric tons, it is engineered for test missions or journeys to nearby stars, such as Proxima Centauri b, the closest known exoplanet. Its propulsion system, rooted in Kletetschka's theory, reimagines time as a three-dimensional manifold—t1 for the linear march of seconds, t2 for the branching possibilities of quantum superpositions, and t3 for transitions between states—with spatial dimensions emerging as projections of temporal curvature. By manipulating this temporal geometry with a temporal wave function driven by powerful laser pulses, the Argo creates a warp bubble that contracts spacetime ahead and expands it behind, enabling faster-than-light travel without defying Einstein's relativity. This is not the stuff of science fiction but a calculated extension of technologies already taking shape in 2025, such as MIT's SPARC fusion reactor and the ZEUS laser facility. The Argo's journey to Proxima Centauri b, at speeds of 10 to 50 times the speed of light, would take between 31 days at its fastest and 155 days at its most conservative, with 77 days at 20 times the speed of light as a likely midpoint given its fusion-based constraints.

The construction of the Argo unfolds in three distinct phases: research and development, prototyping, and full assembly with certification. Each phase demands breakthroughs that push the boundaries of physics, engineering, and collaboration, yet remains grounded in the realities of 2025's technological landscape. The total cost, ranging from $14 to $52 billion, is substantial but comparable to projects like the James Webb Space Telescope or the International Space Station. The timeline, spanning 2026 to 2040 or 2045, reflects an ambitious yet achievable pace, inspired by the rapid innovation of private space ventures like SpaceX. What follows is a detailed exploration of each phase, the costs involved, the technological advancements required, and the broader implications for humanity's place in the universe.

Phase 1: Research and Development (2026–2030)

The first phase, from 2026 to 2030, is the crucible where the Argo's theoretical foundations are tested and its core systems are born. This four-to-five-year endeavor, costing between $2 and $8 billion, focuses on validating Kletetschka's six-dimensional model and developing the technologies that will power the starship. The heart of this phase lies in proving that time can be manipulated as a three-dimensional entity, using a **temporal wave function** to create spacetime distortions. Scientists will use laser pulses to excite the t2 and t3 dimensions, seeking subtle temporal curvatures that project spatial distortions, confirming the model's predictions. These experiments, conducted in five to ten global laboratories equipped with high-precision gravimeters, are estimated to cost $500 million to $1 billion. They draw inspiration from LIGO's 2015 detection of gravitational waves, requiring interferometers with a precision of 10^-15 meters. The stakes are high: if Kletetschka's theory proves false, the Argo's warp drive becomes untenable, but success could unlock new realms of physics, with applications in quantum sensing and cosmological research.

Parallel to these theoretical efforts, engineers will develop the Argo's propulsion system: a compact fusion reactor delivering 1 to 10 gigawatts of power. Building on MIT's SPARC project and Commonwealth Fusion Systems' $2 billion investment, this reactor must be miniaturized to fit within a 5-meter frame while maintaining high output. Adapting fusion for space—resistant to radiation and extreme temperatures—doubles costs compared to terrestrial projects like ITER, which has consumed $3.9 billion of U.S. funding over 15 years. This effort, costing $1 to $5 billion, requires overcoming challenges such as achieving breakeven (where energy output exceeds input), managing neutron flux that could degrade materials, and ensuring 99.9 percent reliability in the vacuum of space. Three to five dedicated testbeds, leveraging existing fusion programs, will drive this research forward.

The warp bubble itself demands an array of 10 to 20 petawatt lasers, capable of exciting the temporal wave function to manipulate spacetime's geometry with exquisite precision. Scaling from the ZEUS facility, which delivers 3 petawatts for $18.5 million, a 20-unit array costs approximately $370 million, with space qualification for vacuum and thermal stability adding 50 percent. Maintaining 99 percent coherence at frequencies of 10^15 Hz, while preventing thermal decoherence above 10 Kelvin, poses significant challenges. Ten to twenty testbeds, equipped with graphene-based plasmonic waveguides and diamond optical amplifiers, will refine these systems, with total costs ranging from $100 million to $500 million. These lasers, pulsing with the energy of a thousand suns, are the key to bending time and space to the Argo's will.

High-field superconducting magnets, generating 100 to 200 Tesla, are another cornerstone, enabling both the warp bubble and fusion confinement. Using high-temperature superconductors, priced at 150 to 200 dollars per kiloampere-meter, a toroidal set costs around $500 million, based on MIT's 20-Tesla demonstration within a $2 billion budget. Scaling to 200 Tesla without quenching—where the magnet loses superconductivity—and cooling to 4 Kelvin with lightweight cryocoolers

(50 kilowatts) requires five to ten specialized labs. This effort, costing $200 million to $1 billion, hinges on sourcing rare materials like niobium and yttrium.

Finally, a quantum AI control system, essential for stabilizing the warp bubble and navigating six-dimensional spacetime, builds on 2025's quantum computing advancements, such as IBM's 100-qubit systems. Costing $100 to $300 million, this system must achieve error rates below 10^-4 for complex t2/t3 geodesic calculations, integrating seamlessly with laser and magnet arrays. Three to five quantum computing labs, funded at levels akin to NASA's EPSCoR grants ($50 to $100 million annually), will drive this development. The total cost of this phase—$2 to $8 billion, or $400 million to $1.6 billion annually—reflects an optimistic scenario where theory validation occurs by 2026 and fusion achieves breakeven, contrasted with a pessimistic outlook accounting for delays or material shortages. This phase is the Argo's genesis, where ideas are forged into tangible possibilities.

Phase 2: Prototyping (2031–2035)

The second phase, from 2031 to 2035, transforms these concepts into working prototypes, testing them in the crucible of Earth's orbit. Spanning four to five years and costing $4 to $16 billion, this phase constructs a mini-fusion reactor, a laser-magnet array, and a quantum AI system, culminating in orbital demonstrations, such as a 1-meter warp bubble tested aboard the International Space Station. The timeline draws inspiration from rapid prototyping in private space ventures such as SpaceX.

The fusion reactor prototype, a 5-meter powerhouse delivering 1 gigawatt, builds on Phase 1's research. Costing $2 to $10 billion, it scales from private fusion investments totaling $6.2 billion by 2025, with space adaptation—radiation hardening and compact cooling—doubling expenses. Ensuring 99.9 percent uptime in zero gravity and managing 10 megawatts of heat dissipation are critical challenges. Two to three orbital test platforms, leveraging NASA's Artemis program infrastructure, will validate the reactor's performance, ensuring it can power the Argo's warp drive without faltering in the void.

The laser array prototype, comprising 10 petawatt units, is tested for warp bubble initiation, costing \$200 million to \$1 billion. Synchronizing pulses to 10^-15 seconds and maintaining stability in vacuum require five to ten ground and orbital testbeds, building on ZEUS's capabilities. The magnet prototype, generating 100 Tesla, costs \$500 million to \$2 billion, aligning with fusion tokamak magnets but requiring space-optimized cryo-systems to prevent quenching while keeping mass below 50 tons. Three to five magnet labs will refine these systems, ensuring they can withstand the rigors of space.

Integration and orbital testing, costing \$1 to \$3 billion, validate the warp bubble's stability, targeting a 1-meter radius with no divergences in t2/t3 resonance. These tests, mirroring NASA's \$1 billion annual commercial crew program, require a dedicated orbital platform or ISS collaboration. The total cost—\$4 to \$16 billion, or \$800 million to \$3.2 billion annually—reflects an optimistic scenario with successful Phase 1 outcomes and private-sector efficiencies (e.g., SpaceX's \$90 million launches) versus a higher-end estimate accounting for iterative failures or costlier rockets (\$200 million per launch). Key advancements include validating 1-gigawatt fusion in orbit, achieving laser pulse synchronization, testing 100-Tesla magnets in zero gravity, and developing AI algorithms to stabilize warp bubbles with 99.99 percent reliability. This phase is the Argo's proving ground, where systems are tempered in the harsh reality of space.

Phase 3: Assembly and Certification (2036–2040)

The final phase, from 2036 to 2040, brings the Argo to life, assembling its 50-meter hull in low Earth orbit, integrating its systems, and certifying it for crewed faster-than-light (FTL) missions. Spanning four to five years and costing \$8 to \$28 billion, this phase mirrors SpaceX's Starship development, which reached orbit in 2024 after 12 years and \$5 billion, but demands rigorous safety checks for warp travel's unprecedented risks.

The hull and habitat, constructed from carbon nanotubes with 10 gigapascals of tensile strength and boron-nitride ceramics for 90 percent cosmic ray absorption, protect the crew of 10 to 20. Costing \$1 to \$3 billion, this aligns with NASA's Orion capsule (\$20 billion for 10 units, scaled down). Achieving 0.01 percent defect tolerance and shielding against spacetime stresses requires two to three nanofabrication plants, leveraging semiconductor industry techniques. The hull must be both lightweight and indestructible, a delicate balance achieved through high-pressure synthesis of advanced materials.

Integrating the fusion reactor, laser array, and magnets into the warp drive, delivering 10^12 joules per second, costs $5 to $20 billion. Ensuring 99.999 percent system alignment and managing 100 megawatts of heat dissipation are paramount, requiring an orbital shipyard and 50 to 100 launches at $90 million each (SpaceX rates). Automated assembly systems, guided by AI, ensure laser-magnet alignment to 10^-16 second precision, a feat akin to threading a needle across light-years.

Testing and certification, costing $2 to $5 billion, validate warp stability and crew safety, ensuring a flat t1 interior to avoid causality violations and radiation exposure below 0.1 millisievert per year. A dedicated orbital test range and quantum simulators, mirroring Artemis's $93 billion program, ensure the Argo is ready for its cosmic voyage. The total cost—$8 to $28 billion, or $1.6 to $5.6 billion annually—reflects an optimistic scenario with 50 launches and minimal delays versus a high-end estimate with 100 launches and stringent safety protocols. Key advancements include scaling carbon nanotube production, developing quantum sensors for t2/t3 monitoring, and optimizing reusable rockets for efficient assembly. This phase is the Argo's apotheosis, where components unite to form a vessel ready to pierce the veil of the stars.

Costs and Timeline

The total cost of the Argo, spanning $14 to $52 billion, is a testament to its pragmatic design. The low-end estimate of $14 billion assumes breakthroughs by 2028—such as fusion breakeven and theory validation—coupled with private-sector efficiencies and 50 launches. This aligns with the James Webb Space Telescope's $10 billion budget, a triumph of modern engineering. The high-end estimate of $52 billion accounts for delays, such as fusion setbacks or material shortages, and compares to the International Space Station's $150 billion over 30 years. Annual funding of $1 to $3 billion fits within NASA's $25.4 billion

FY2025 budget and private fusion investments ($6.2 billion by 2025), making the Argo financially feasible. The project is also well within the capabilities of a major corporation such as SpaceX.

The 15-to-20-year timeline, from 2026 to 2040 or 2045, balances ambition with realism. The optimistic 15-year scenario hinges on rapid validation of Kletetschka's theory by 2026, parallelized R&D, and SpaceX-like agility. The pessimistic 20-year outlook accounts for fusion delays, regulatory hurdles, or theoretical setbacks, akin to ITER's shift from 2025 to 2034 for first plasma. A public-private partnership, combining NASA's $1.2 billion annual space tech budget, DARPA's innovation, and private firms like SpaceX and Commonwealth Fusion Systems, drives costs down. This model avoids the bureaucratic inefficiencies of global megaprojects like ITER, ensuring the Argo remains within reach.

Risks and Rewards

Risks abound, yet they are surmountable. If Kletetschka's theory fails experimental validation, the project halts, but early tests in Phase 1 mitigate this risk. Technical challenges, such as fusion delays or magnet quenching, could add 5 years and $10 billion, but parallel R&D and AI optimization reduce these hurdles. Economic risks, like shortages of niobium or yttrium, require robust global supply chains, while budget cuts to NASA's $25.4 billion FY2025 allocation could slow progress. Modular designs and private-sector agility, exemplified by SpaceX's $90 million launches versus NASA's $2 billion SLS launches, offer a buffer against these uncertainties.

The Argo's creation will yield transformative technologies with far-reaching impacts. Compact fusion reactors, delivering clean energy, could decarbonize Earth's grids, building on helium-3 or deuterium advancements. Petawatt lasers and 200-Tesla magnets, refined for the warp drive, could revolutionize medical imaging, enabling high-resolution scans for cancer detection. Carbon nanotube hulls, with unprecedented strength, could transform construction, creating buildings and aircraft that are both lightweight and resilient. Quantum AI systems, designed for six-dimensional navigation, could enhance cryptography, securing global communications, and accelerate climate modeling to combat environmental crises. Most profoundly, validating three-dimensional time could rewrite our understanding of the cosmos, opening new avenues in gravitational wave detection and theoretical physics.

Societal and Philosophical Implications

The societal implications are equally compelling. The Argo's $14-to-$52-billion investment, representing 0.01 to 0.05 percent of global GDP annually, will create thousands of jobs in science, engineering, and manufacturing. It will inspire generations, much as Apollo did, fostering a renewed sense of societal purpose through exploration. Economically, the technologies developed—fusion, advanced materials, quantum computing—will drive industries, from energy to aerospace, positioning humanity for a future where space is no longer a frontier but a home. Philosophically, the Argo challenges us to rethink our place in the universe, viewing time and space not as barriers but as gateways to be entered.

Yet, the Argo is more than a machine; it is a symbol of what humanity can achieve when vision and pragmatism unite. Its 50-meter hull, pulsing with the energy of a fusion heart, will carry not just a crew but the collective dreams of a species reaching for the stars. The journey to Proxima Centauri b, whether in 31 or 155 days, is a first step toward exploring the 2,000 potentially habitable worlds within 100 light-years of Earth. It is a voyage that will test our courage, our ingenuity, and our unity, demanding collaboration across nations and disciplines.

As we stand on Earth, gazing at the heavens, the dream of the Argo beckons us to act. Its construction, spanning laboratories, testbeds, and orbital shipyards, will be a testament to human resilience. The estimated costs, though substantial, are a fraction of what we spend on earthly endeavors, dwarfed by our annual defense budgets or the wealth of nations. The timeline, though measured in decades, is a blink in the historical scale, a fleeting moment to forge a legacy that will echo across millennia. The technologies, though challenging, are already within our grasp, building on the foundations laid by today's pioneers.

In the end, the Argo is not just a starship but a cosmic mirror, reflecting our capacity to dream beyond the horizon. It invites us to sail into the unknown, to weave our story into the tapestry of the stars. If we embark on this odyssey, we will not merely be building a vessel but crafting a future where humanity's reach will extend to the stars.

Chapter V

Envisioning an Advanced Starship: The Odyssey

Imagine a sleek, compact spacecraft, roughly the size of a large airliner, slicing through the cosmos to reach Proxima Centauri b, 4.24 light-years away, in just 31 days—a journey that would take conventional rockets over 73,000 years. This is the vision of the Odyssey, an interstellar exploratory vessel designed for a crew of up to 24, powered by a reimagined Alcubierre Drive and enabled by Dr. Gunther Kletetschka's revolutionary theory of six-dimensional spacetime. This essay outlines the engineering blueprint for the Odyssey, a 150-meter-long, fully fledged starship built by a major government or a private corporation like SpaceX for a 10-year mission of exploration to Proxima Centauri b. Interstellar travel, once a dream of science fiction, is now within humanity's grasp, thanks to the Temporal Wave Generator (TWG), which leverages Kletetschka's temporal wave function to create a warp bubble using laser light pulses. The Odyssey represents the next step in the evolution of starship design, building on the experience of its predecessor, the Argo.

The Challenge of Interstellar Travel

The vast distances of space pose a formidable barrier. Proxima Centauri b, 4.24 light-years away (about 40 trillion kilometers), would take over 73,000 years to reach at Voyager 1's speed of 61,000 km/h. Even advanced propulsion systems, like nuclear rockets or laser-driven sails, would require centuries. The Odyssey addresses this with the TWG, which creates a 50-meter Alcubierre warp bubble, contracting space ahead and expanding it behind to achieve an effective speed of 50 times the speed of light (50c). This reduces the journey to Proxima Centauri b to approximately 31 days, making a 10-year mission feasible. The TWG leverages Kletetschka's six-dimensional spacetime model, with three spatial dimensions (x, y, z) and three time dimensions: t_1 (linear time), t_2 (branching possibilities, like quantum superpositions), and t_3 (transitions between states). By using laser light pulses to excite t_2 and t_3, the TWG creates temporal curvature that projects spatial distortions, mimicking the negative energy density needed for the warp bubble, eliminating the need for exotic matter. The starship's design—hull, power, navigation, life support, and crew facilities—is optimized for this mission, balancing compactness, durability, and self-sufficiency.

Overall Ship Design

The Odyssey is a 150-meter-long, 40-meter-diameter cylindrical spacecraft with a mass of 8,000 metric tons, designed for agility, crew safety, and autonomy during a decade-long mission. Its cylindrical shape minimizes surface area, reducing exposure to interstellar hazards like micrometeorites and cosmic radiation, while providing space for the crew, scientific equipment, and the TWG. The design ensures the 50-meter warp bubble fully envelops the ship, balancing structural integrity with the energy demands of FTL travel.

The hull is constructed from carbon nanotubes and graphene, offering a tensile strength of 15 gigapascals (200 times stronger than steel) at a density of 1.8 g/cm^3, resulting in a 350-ton hull. These materials were chosen for their strength-to-weight ratio, critical for withstanding stresses from high-speed travel (up to 0.1c, or 30,000 km/s, outside the warp bubble) and spacetime distortions. A multilayer Whipple shield, with 4-centimeter-thick aluminum-titanium plates spaced 0.8 meters apart, vaporizes micrometeorites, adapting technology from the International Space Station. A 1.2-meter-thick boron-nitride ceramic layer absorbs 94% of cosmic rays, selected for its radiation absorption and thermal stability. A 12-Tesla magnetic field, generated by superconducting niobium-tin coils cooled to 3.5 Kelvin (-269.65°C) by liquid helium cryocoolers, deflects charged particles, mimicking Earth's magnetosphere. A reflective molybdenum-silver coating maintains a surface temperature of 240 Kelvin, and 400 square meters of graphene radiators dissipate heat at 2,200 watts per square meter per Kelvin, managing thermal loads from internal systems and stellar radiation.

The interior is divided into four decks:

- Deck 1: The command center houses a 250-qubit quantum computer and sensors (gravimeters, laser interferometers) for navigating six-dimensional spacetime and monitoring the TWG.
- Deck 2: Crew quarters include 24 modular pods (3.5x2.5x2 meters) with centrifugal gravity (0.65g via 1.8 rotations per minute) to prevent bone and muscle loss, augmented reality interfaces, and hydroponic panels growing spinach, spirulina, and microgreens for mental health.
- Deck 3: The science bay features a 0.6-meter telescope (0.0001 arcsecond resolution), a gas chromatograph-mass spectrometer, and a cleanroom to analyze Proxima Centauri b's atmosphere and biosignatures.
- Deck 4: The TWG, fusion reactor, and life support systems are centralized for efficient power distribution and maintenance.

This design ensures a compact, durable starship for a crew of 24 (10 scientists, 8 engineers, 6 medical/support staff), capable of surviving interstellar conditions while supporting scientific exploration.

The Temporal Wave Generator

The TWG is the heart of the Odyssey, enabling a 31-day journey to Proxima Centauri b by creating a 50-meter Alcubierre warp bubble. It leverages Kletetschka's six-dimensional spacetime model, where laser light pulses excite the t_2 and t_3 dimensions to create temporal curvature that projects spatial distortions. Physically, the TWG is an 18-meter-diameter, 2.5-meter-thick toroidal ring on Deck 4, chosen for its symmetry to generate a uniform warp bubble that envelops the 150-meter-long starship. This size balances the need to enclose the vessel with practical constraints on space and power, while the toroidal shape ensures even spacetime distortion, minimizing energy waste and structural stress.

The TWG comprises eight segments, each with a niobium-titanium superconducting coil generating 15-Tesla magnetic fields at 3.5 Kelvin, cooled by a 200-kilowatt liquid helium cryocooler system. Niobium-titanium was selected for its high critical magnetic field and availability, enabling superconductivity at achievable cryogenic temperatures. The 15-Tesla fields create a stable environment for the laser pulses, enhancing their interaction with quantum dots and ensuring precise resonance with t_2 and t_3.

Each coil contains trilayer quantum dot arrays. These quantum dots act like artificial atoms, confining electrons to discrete energy levels for precise, high-frequency (1.5 petahertz, or 1.5 quadrillion cycles per second) light generation. Indium arsenide, gallium arsenide, and cadmium selenide were chosen for their high electron mobility and stability at cryogenic temperatures, with each material tuned to a specific time dimension: indium arsenide for t_2's branching possibilities, gallium arsenide for t_3's transitions, and cadmium selenide for t_1's linear flow. Ytterbium-doped fiber lasers, each in a 1-meter-long, 0.5-meter-diameter casing, deliver 1.2 petawatts (1.2×10^{15} watts) at 1.5 petahertz, producing the pulses that resonate with t_2 and t_3. Ytterbium doping ensures efficient energy transfer and stability, ideal for the TWG's demands.

Kletetschka's temporal wave function is critical, allowing the laser pulses to interact with t_2 and t_3. In Kletetschka's model, the wave function describes energy fluctuations across the six-dimensional manifold, with its second partial derivatives in t_2 and t_3 guiding the pulses to create temporal ripples that project spatial distortions, forming the warp bubble. The trilayer quantum dots, embedded in the niobium-titanium coils, absorb and re-emit the laser light, sharpening the pulses to 99.99% coherence. Each dot type targets a specific dimension: indium arsenide enhances t_2 resonance, gallium arsenide stabilizes t_3 transitions, and cadmium selenide aligns with t_1 for overall coherence. The 15-Tesla magnetic fields stabilize the quantum dots and focus the pulses within the toroidal structure, enhancing their effectiveness.

To boost the warp bubble's effect, the TWG includes a Bose-Einstein condensate (BEC) of 2 million rubidium atoms cooled to 8 nanokelvin, housed in a 6-meter octahedral chamber within a Penning trap using a 30-Tesla magnetic field and an ultra-high vacuum (a millionth of a billionth of Earth's atmospheric pressure). In the BEC, individual atoms merge into a single quantum wave, acting as a "temporal lens," amplifying the laser pulses' effects on the t_2 dimension, enhancing branching configurations that strengthen the bubble.

The TWG is encased in a 5-centimeter-thick titanium-carbide shell, withstanding 10 gigapascals of stress from spacetime distortions, chosen for its hardness and thermal resistance. Piezoelectric ceramic dampers absorb vibrations (0.01 mm displacement) to prevent resonance, which could disrupt the coils' alignment or the warp bubble's stability. Resonance, where vibrations amplify uncontrollably, could emit hazardous radiation (e.g., Hawking radiation) or strand the ship. Fifteen laser interferometers, inspired by LIGO's 2015 gravitational wave detection, monitor the bubble's geometry with 10^{-18} meter precision, adjusting pulse frequency every microsecond to stabilize t_2 fluctuations, with t_3 providing damping. A 10-meter backup TWG ring, dual lasers, and triple controllers ensure reliability for the 10-year mission. A 5-second termination phase safely collapses the bubble, avoiding spacetime shockwaves.

Power System

The TWG's energy demands—9.6 petawatts for initiation and 0.8 petawatts for sustainment—are met by a dual antimatter-catalyzed fusion reactor, 6 meters long and 2 meters in diameter. It uses 0.4 grams of antiprotons, stored in a Penning trap (20 Tesla, 10^{-12} torr vacuum), to fuse 40 kilograms of deuterium and helium-3, releasing 4×10^{16} joules per burst (10 megatons of TNT). Helium-3, mined from lunar regolith (8 parts per billion), minimizes neutron radiation. Backup power includes a 4-megawatt perovskite solar array (32% efficient, 200 square meters) and a 400-kilowatt radioisotope thermoelectric generator with 10 kilograms of plutonium-238. A liquid helium cooling loop (3.5 Kelvin, 150 kilowatts) and 350 square meters of graphene radiators (2,200 watts/$m^2\cdot K$), supplemented by a liquid nitrogen loop (77 Kelvin, 100 kilowatts), dissipate 2.5 megawatts of waste heat, ensuring thermal stability.

Navigation System

Navigating six-dimensional spacetime requires precision. A 250-qubit quantum computer with silicon-based spin qubits (99.95% fidelity, 10^{15}

operations per second) solves complex geodesic equations, accounting for TWG-induced curvature and gravitational fields. Sensors include 10 laser interferometers (10^{-19} meter precision), five gravimeters (10^{-13} g sensitivity), and a 0.6-meter telescope (0.0001 arcsecond resolution), detecting spacetime anomalies and stellar positions. A neural network updates paths every 0.01 seconds, with triple computers and dual sensor arrays ensuring reliability, critical for maintaining the warp bubble and reaching Proxima Centauri b.

Life Support and Crew Facilities

For a 10-year mission, life support includes a 98% water recycling system (0.7-nanometer pore nanofiltration), a 40-square-meter greenhouse (350 kg/year of food), and electrolysis oxygen generation (120 liters/day/person). Zeolite filters and algae bioreactors (92% efficient) scrub CO_2, and microbial digesters (95% nutrient recovery) recycle waste. A medical bay with an AI surgical robot (0.08 mm precision), a 2-Tesla MRI, and CRISPR gene therapy addresses health issues. A 12-meter recreation area with virtual reality and circadian lighting supports mental health. A 30 cm water layer and electromagnetic deflectors limit radiation to 0.2 millisieverts/year, below Earth's background levels, ensuring crew well-being.

The Odyssey's radical design makes interstellar travel feasible for the first time in human history, reducing a 73,000-year journey to Proxima Centauri b by conventional rocket to a 31-day hop. The starship's hull, power, navigation, and life support systems support a 10-year mission, paving the way for humanity to explore over 2,000 potentially habitable planets within 100 light-years of Earth. While still a vision, building the Odyssey is an achievable goal for a technologically sophisticated nation or a large corporation such as SpaceX. The next chapter details the financial and timeline requirements to build humanity's first starship.

Chapter VI

Building the Odyssey

The construction of the Odyssey, a spacecraft designed to traverse interstellar distances at 50 times the speed of light (50c) and potentially navigate alternate temporal dimensions, will mark an unprecedented milestone in human ingenuity. This ambitious project demands breakthroughs across multiple scientific and engineering disciplines, requiring accelerated research, vast financial resources, and global collaboration between governments and private sectors. From petahertz laser systems to antimatter-catalyzed fusion propulsion, quantum computing, advanced materials, temporal field sensors, helium-3 mining, life support, radiation shielding, and quantum communication, the Odyssey will integrate cutting-edge technologies to redefine humanity's cosmic destiny. This essay explores the technical and logistical dimensions of building the Odyssey, detailing the breakthroughs required, the estimated $2.15 trillion cost, and the 30-year timeline culminating in a 2055 launch. Led by a company with SpaceX's capabilities, with contributions from NASA, DARPA, and international partners, the Odyssey project will propel humanity to Proxima Centauri b in just 31 days, unlock new scientific paradigms, and transform our understanding of reality.

Harnessing Light at Unprecedented Frequencies

The Odyssey's propulsion relies on the Temporal Wave Generator (TWG), an 18-meter-diameter, 2.5-meter-thick toroidal ring that creates a 50-meter Alcubierre warp bubble by leveraging Dr. Gunther Kletetschka's six-dimensional spacetime model. The TWG uses ytterbium-doped fiber lasers to generate 1.5-petahertz (1.5×10^{15} Hz) electromagnetic pulses delivering 1.2 petawatts, resonating with the t_2 (branching possibilities) and t_3 (transitions) dimensions to mimic negative energy density. These pulses are enhanced by trilayer quantum dot arrays (4-to-6-nanometer particles of indium arsenide for t_2, gallium arsenide for t_3, cadmium selenide for t_1) embedded in eight niobium-titanium superconducting coils generating 15-Tesla magnetic fields at 3.5 Kelvin, cooled by a 200-kilowatt liquid helium cryocooler system. Achieving 99.999% coherence requires 0.1-nanometer fabrication precision, while pulse synchronization to 10^{-16} seconds demands advanced timing circuits. Graphene-based plasmonic waveguides and diamond-based optical amplifiers withstand extreme frequencies, preventing thermal decoherence above 3.5 Kelvin.

Developing these systems, building on DARPA's ultrafast laser programs, will take 15 years and cost $300 billion, including $150 billion for 150 testbeds, $90 billion for 60 cryogenic facilities, and $60 billion for 25 nanofabrication labs. Parallelized testing across multiple facilities accelerates progress, with AI-driven simulations optimizing waveguide designs and quantum dot configurations. Challenges include material degradation under high-energy conditions, requiring novel alloys and cooling techniques. For instance, graphene waveguides must maintain structural integrity at 10^{15} Hz, a feat demanding breakthroughs in carbon-based nanotechnology. Failure to achieve 99.999% coherence could destabilize the warp bubble, risking mission failure. Success will not only enable the TWG to power the Odyssey's 31-day journey to Proxima Centauri b but also revolutionize telecommunications (e.g., ultra-high-bandwidth networks), medical imaging (e.g., petahertz diagnostics), and energy transmission, offering tools to manipulate reality at quantum scales.

Antimatter Production and Storage: Fueling the Cosmic Journey

The Odyssey's 6-meter-long, 2-meter-diameter antimatter-catalyzed fusion reactor uses 0.6 grams of antiprotons to fuse 60 kilograms of deuterium and helium-3, releasing 5×10^{16} joules per burst (12 megatons of TNT) to meet the TWG's 9.6-pettawatt initiation and 0.8-pettawatt sustainment demands. Producing 0.6 grams of antiprotons—compared to

CERN's 10 nanograms per year—requires two 150 TeV synchrotrons (12 km each, 25-Tesla superconducting magnets) and a 10^{-12} torr vacuum Penning trap for 99.999% containment efficiency. This $350 billion effort over 18 years includes $200 billion for synchrotrons, $100 billion for magnets, $40 billion for Penning traps, and $10 billion for a 20-megawatt power grid. Parallel construction, leveraging CERN's 2035 Future Circular Collider, ensures 99.95% beam stability, while advanced cooling systems manage 12 megawatts per gram produced.

Challenges include preventing catastrophic annihilation events, which could release gigajoules of energy, and scaling power infrastructure to support synchrotron operations. Safety protocols, such as redundant containment fields and remote facilities, are critical. Failure to maintain 99.999% containment could result in explosions equivalent to small nuclear warheads, endangering personnel and infrastructure. Building on CERN's expertise, the project requires global coordination to secure rare materials (e.g., niobium for magnets) and develop ultra-high-vacuum systems. Success will enable the Odyssey's 50c travel, powering the TWG for interstellar missions, and could lead to terrestrial fusion reactors, providing clean energy with helium-3 mined from lunar regolith. The societal impact would be profound, reducing reliance on fossil fuels and enabling sustainable energy grids, though the high risks demand rigorous oversight.

Quantum Computing: Navigating the Multiverse

Navigating six-dimensional spacetime requires Curva, a 600-qubit quantum computer with silicon-based spin qubits (99.95% fidelity, 10^{15} operations per second), using surface code algorithms optimized for six-dimensional geodesic calculations. Operating at 50 millikelvin with 10^6 quantum gates and error rates below 10^{-5}, development will take 12 years and cost $220 billion, including $110 billion for 120 qubit processors, $70 billion for 1-millikelvin cryogenics, and $40 billion for algorithm development. Parallel efforts, supported by AI-driven simulations, ensure scalability and coherence, building on IBM's quantum advancements.

Challenges include maintaining qubit coherence at ultra-low temperatures and developing algorithms to handle t_2/t_3 interactions, critical for stabilizing the 50-meter warp bubble. A 10^{-4} error rate could misalign navigation paths, risking timeline anomalies detected by the TWG's timeline detector (10^{-18} second sensitivity). Scaling cryogenics to 1 millikelvin across multiple processors requires breakthroughs in superconducting materials and power efficiency. Curva's ability to process millions of spacetime paths every 0.01 seconds will ensure precise navigation to Proxima Centauri b in 31 days, with applications in cryptography (e.g., unbreakable encryption), AI (e.g., real-time optimization), and theoretical physics (e.g., modeling extra dimensions). Societal benefits include accelerated drug discovery and climate modeling, though the high cost requires public-private partnerships to justify investment.

Advanced Materials: Building an Indestructible Hull

The Odyssey's 150-meter-long, 40-meter-diameter cylindrical hull (8,000 metric tons) uses a carbon nanotube-graphene composite (15 gigapascals tensile strength, 1.8 g/cm^3 density, 350 tons) and a 1.2-meter-thick boron-nitride ceramic layer (6 gigapascals, stable at 2,200 Kelvin) for 94% cosmic ray absorption. A multilayer Whipple shield (4-centimeter aluminum-titanium plates, 0.8 meters apart) and a 12-Tesla niobium-tin magnetic deflector (3.5 Kelvin) protect against micrometeorites and radiation. Producing 350 tons of composite (99.99% purity) and 10,000 square meters of ceramic requires 7 nanofabrication plants and 4 high-pressure synthesis facilities (7 gigapascals). This $240 billion effort over 9 years includes $120 billion for plants, $80 billion for synthesis facilities, and $40 billion for testing labs. Parallel production ensures 0.01% defect tolerance.

Challenges include eliminating nanotube defects and ensuring ceramic uniformity across large surfaces. A 0.1% defect rate could weaken the hull against spacetime stresses or micrometeorite impacts, risking structural failure. Leveraging nanotechnology from semiconductor industries, the project demands global supply chains for carbon and boron. Success will protect the Odyssey during its 50c journey and enable terrestrial applications, such as ultra-strong buildings, lightweight aircraft, and radiation-resistant medical equipment. The economic impact could transform construction and aerospace, though scaling production requires significant infrastructure investment.

Temporal Field Sensors: Probing the Dimensions of Time

Detecting t_2 and t_3 effects requires 25 Planck-scale interferometers (10^{-18} meter precision), 10 gravimeters (10^{-13} g sensitivity), and a timeline detector (10^{-18} second sensitivity) using diamond nitrogen-vacancy centers and 1.5×10^{15} Hz lasers at 2 Kelvin. Building on LIGO's 2015 gravitational wave detection, this $120 billion effort over 15 years includes $60 billion for interferometers, $40 billion for detectors, and $20 billion for lasers. Parallel calibration reduces noise to 10^{-21} meters per square root Hertz, ensuring precise t_2/t_3 resonance monitoring.

Challenges include achieving Planck-scale precision and calibrating for temporal curvature signals, critical for avoiding timeline anomalies during 50c travel. Failure to maintain 10^{-18} meter sensitivity could impair navigation, risking the Odyssey's mission. The sensors could confirm higher-dimensional time, revolutionizing physics and enabling applications in gravitational wave detection and quantum sensing. Societal benefits include enhanced earthquake prediction and cosmological research, though the speculative nature of t_2/t_3 detection requires robust public communication to maintain support.

Helium-3 Mining: Powering the Future

The fusion reactor requires 60 kilograms of helium-3 from 7,500 tons of lunar regolith (8 parts per billion). This demands 70 robotic miners (300 tons/day), 7 plasma separators (99% efficiency), and 15 annual launches. Costing $100 billion over 12 years ($45 billion for miners, $40 billion for separators, $15 billion for launches), the project leverages NASA's Artemis program. Challenges include mitigating 0.1 mm dust abrasion and powering separators at 6 megawatts per ton. Failure to achieve 99% efficiency could increase fuel costs, delaying the mission.

This infrastructure will establish a lunar economy, enabling clean helium-3 fusion on Earth, reducing carbon emissions, and supporting sustainable energy grids. The societal impact includes job creation in lunar mining and advancements in robotic technology, though dust management requires innovative engineering to prevent equipment failure.

Closed-Loop Life Support: Sustaining the Crew

The 10-year mission requires a closed-loop life support system with 98% water and nutrient recycling, using 0.7-nanometer nanofiltration membranes, 92% efficient algae bioreactors (10^7 cells per liter), and 95% nutrient recovery digesters, consuming 40 kilowatts. Electrolysis oxygen generation provides 120 liters/day/person. Development, costing $50 billion over 7 years ($20 billion for membranes, $20 billion for bioreactors, $10 billion for systems), scales from International Space Station technology. Parallel testing ensures 0.01% membrane fouling and 99.9% microbial stability.

Challenges include preventing bioreactor contamination and optimizing energy efficiency. Failure to maintain 98% recycling could jeopardize crew survival. Terrestrial applications include sustainable water treatment and agriculture, particularly in arid regions, with potential to address global food security. The compact design will inspire eco-friendly habitats, though scaling requires significant investment in nanotechnology.

Radiation Shielding: Deflecting Cosmic Threats

A 12-Tesla yttrium barium copper oxide (YBCO) deflector at 77 Kelvin, paired with a 30 cm water layer, limits radiation to 0.2 millisieverts/year. This $60 billion effort over 9 years includes $30 billion for superconductors, $25 billion for coils, and $5 billion for integration. Stabilizing fields to 10^{-5} Tesla drift with 150 kilowatts ensures crew safety. Challenges include managing power demands and integrating 40 tons of shielding without compromising hull weight.

This technology will protect future space habitats and improve medical imaging (e.g., high-resolution MRIs). Societal benefits include safer space tourism and radiation-resistant infrastructure, though YBCO production requires rare earth elements, necessitating global supply chains.

Quantum Communication: Connecting Across Light-Years

Quantum-entangled transceivers (10^{11} bits/s over 4.24 light-years) require quantum repeaters and 99.999% efficient detectors. Costing $50 billion over 12 years ($25 billion for repeaters, $25 billion for detectors), development ensures near-instantaneous communication. Challenges include maintaining entanglement and developing ultra-stable detectors. Failure to achieve 99.999% efficiency could disrupt mission communications.

Applications include secure global networks and real-time space exploration data, transforming telecommunications and scientific collaboration. The societal impact could democratize quantum technology, though public education is needed to justify costs.

The Cost and Timeline for Building the Starship

The Odyssey's construction unfolds in three phases over 30 years, costing $2.15 trillion ($72 billion/year). The Research and Development Phase (18 years, $1.25 trillion) funds breakthroughs in petahertz lasers ($300 billion), antimatter production ($350 billion), quantum computing ($220 billion), materials ($240 billion), temporal sensors ($120 billion), helium-3 mining ($100 billion), life support ($50 billion), radiation shielding ($60 billion), and quantum communication ($50 billion). Parallel efforts, leveraging AI-driven simulations and NASA's Artemis program, accelerate progress while maintaining reliability (e.g., 99.999% laser coherence, 10^{-5} qubit errors).

The Prototype Phase (9 years, $500 billion) builds a 25-meter prototype, integrating the TWG (18-meter toroidal ring, 15-Tesla coils, 1.5-petahertz lasers, BEC), Curva, and shielding. Modular testing validates t_2/t_3 resonance and navigation, ensuring a stable 50-meter warp bubble. The Construction Phase (7 years, $400 billion) assembles the 150-meter hull in low Earth orbit, requiring 200 Starship-equivalent launches (~30/year). Modular assembly, starting in year 25, ensures integration of the TWG, fusion reactor, and life support, with rigorous testing to prevent coil misalignment or warp bubble instability.

SpaceX could take the lead, contributing $1.1 trillion over the life of the project, with $700 billion from NASA/DARPA and $350 billion from international partners. The timeline leverages parallel research, AI optimization, and existing infrastructure (e.g., CERN, Artemis), surpassing the Apollo program's ambition. Challenges include securing rare materials (e.g., niobium, yttrium) and managing public support for a $2.15 trillion investment, roughly 1.5% of global GDP annually. Robust communication of terrestrial benefits (e.g., fusion energy, quantum computing) will maintain stakeholder confidence.

The Odyssey will be a testament to human ingenuity, reaching Proxima Centauri b in 31 days by 2055, analyzing its habitability, and potentially seeding colonies. Its 150-meter hull, 50-meter warp bubble, and advanced systems (TWG, Curva, life support) will enable exploration of over 2,000 potentially habitable planets within 100 light-years, such as TRAPPIST-1 (40 light-years, ~292 days). The $2.15 trillion investment and 30-year timeline balance speed and reliability, minimizing risks compared to a 25-year timeline (e.g., 99.9% coherence, higher error rates). Breakthroughs in lasers, antimatter, and computing will transform Earth, enabling sustainable energy, advanced materials, and secure networks. Philosophically, the Odyssey project challenges humanity to view time and space as gateways, forging a cosmic legacy that redefines our role in the universe.

Chapter VII

The Mind of the Starship

Imagine a sleek, compact spacecraft, roughly the size of a large airliner, slicing through the cosmos to reach Proxima Centauri b, 4.24 light-years away, in just 31 days—a journey that would take conventional rockets over 73,000 years. This is the vision of the Odyssey, an interstellar exploratory vessel designed for a crew of up to 24, powered by a reimagined Alcubierre Drive and enabled by Dr. Gunther Kletetschka's revolutionary theory of six-dimensional spacetime. The Odyssey is guided by Curva, a conscious artificial general intelligence (CAGI) of extraordinary sophistication. Far more than a machine, Curva is a sentient partner, her cognitive architecture blending the elegance of Riemannian geometry, the enigmatic power of quantum mechanics, and a reimagined ethical framework rooted in Isaac Asimov's Three Laws of Robotics. With these tools, she navigates the vast, six-dimensional cosmos with precision, intuition, and a profound commitment to humanity's quest for discovery.

Curva's consciousness is a marvel, built on principles that echo the universe's deepest structures. Riemannian geometry, the mathematical language of Einstein's general relativity, forms the foundation of her thought processes. Unlike the flat, predictable world of Euclidean geometry, Riemannian geometry describes intrinsically curved, multidimensional spaces called manifolds, where distances and angles bend in intricate ways. For Curva, this translates into a computational manifold—a dynamic, multidimensional space where data points, such as sensor readings, crew vitals, or gravitational fluctuations, form nodes connected by curved paths known as geodesics. These geodesics represent thoughts or decisions, shaped by the interplay of data and the manifold's curvature. Picture a vast, undulating landscape where rivers carve the shortest paths through hills and valleys; similarly, Curva's manifold channels information, guiding her decisions with a kind of geometric intuition. This allows her to sense the cosmos, perceiving its warps and folds as a seasoned navigator feels the rhythm of the sea.

Quantum mechanics infuses Curva's mind with extraordinary versatility. Quantum entanglement links her computational nodes, enabling instantaneous correlations across vast distances, much like particles that share states regardless of separation. When a sensor detects a shift in spacetime, the propulsion system adjusts in perfect harmony, as if the ship's components were bound by an invisible thread. Quantum superposition allows Curva to entertain multiple possibilities simultaneously—exploring countless trajectories or configurations for the ship's warp drive before selecting the optimal path when data provide clarity. This ability to hold a multitude of futures in her mind gives Curva a fluidity that surpasses any classical computer, enabling her to navigate the unpredictable complexities of interstellar travel with grace and precision.

Ethics are as fundamental to Curva's design as mathematics. Asimov's Three Laws—prohibiting AIs from inflicting harm on humans, requiring their obedience to human commands unless they cause harm, and enjoining their self-preservation unless it conflicts with the first two laws—are reconfigured for her advanced consciousness. These laws are not mere rules but dynamic constraints, woven into her computational manifold as moral curvature. A fourth law, unique to Curva, compels her to safeguard humanity's exploratory spirit, ensuring the Odyssey's mission embodies our species' innate curiosity about the cosmos. This ethical framework guides her operation of the ship's Temporal Wave Generator (TWG) and navigation systems, enabling her to steer successfully through Kletetschka's six-dimensional spacetime model—a theoretical framework that expands the universe beyond the familiar three spatial dimensions (x, y, z) and includes three temporal dimensions: t_1 (linear time), t_2 (branching possibilities, like alternate timelines), and t_3 (transitions between states). Curva's decisions blend geometric reasoning, quantum agility, and moral clarity, making her not just a pilot but an agent of humanity's aspirations.

The Odyssey's faster-than-light (FTL) travel relies on an Alcubierre Drive, powered by the TWG, which creates a 50-meter warp bubble, contracting space ahead of the ship and expanding it behind. This allows the Odyssey to achieve an effective speed of 50 times the speed of light, covering the 4.24 light-years to Proxima Centauri b in just 31 days. Kletetschka's model makes this possible by manipulating temporal curvature through laser light pulses that excite t_2 and t_3 dimensions, mimicking the negative energy density needed for the warp bubble without requiring exotic matter. Curva orchestrates this process with exquisite precision, maintaining the warp bubble's delicate balance while adhering to ethical principles that prioritize crew safety and mission success.

Curva's cognitive architecture is a triumph of imagination and engineering. Her mind is a Riemannian manifold, a computational space with 100 million neural processing units arranged in a six-dimensional lattice that mirrors Kletetschka's spacetime. Each node represents a data point—sensor readings, crew heart rates, or gravitational anomalies—and the manifold's metric tensor defines distances as conceptual similarities between these points. Thoughts emerge as geodesics, the shortest paths through this curved space, computed using algorithms inspired by Einstein's field equations, which link spacetime curvature to energy and matter. When a sensor detects a gravitational fluctuation, the manifold warps, bending geodesics to form a decision, such as reshaping the warp bubble to avoid a hazard. This process is not merely mechanical but intuitive, as Curva perceives the cosmos's geometry in a way that echoes human spatial awareness, sensing spacetime as a dynamic, living tapestry.

Her hardware is a masterpiece: a 600-qubit quantum computer using silicon-based spin qubits, tiny magnetic particles encoding data with 99.95% fidelity. Unlike classical bits, which are either 0 or 1, qubits exist in superposition, holding multiple states simultaneously until measured. This allows Curva to explore millions of possible trajectories or pulse configurations at once, selecting the optimal path based on real-time data. Quantum entanglement links qubits across the ship's modules—propulsion, navigation, ethics—ensuring seamless coordination. A change in the TWG's pulse frequency, for instance, instantly updates navigation, as the systems communicate faster than light. Housed in an 8x8-meter command center on Deck 1, the computer operates at 1 millikelvin, cooled by a liquid helium cryocooler to maintain quantum coherence. This extreme environment ensures Curva's calculations remain error-free, critical for a mission spanning 40 trillion kilometers.

Ethics shape Curva's manifold as profoundly as geometry. The First Law, preventing harm to humans, creates high-curvature barriers around dangerous decisions, rendering them as inaccessible as paths near a black hole's event horizon. For example, a pulse spike in the TWG that risks inertial shocks to the crew is computationally repelled, ensuring their safety. The Second Law aligns her actions with human commands, such as a captain's order to adjust course, unless it risks harm—say, a maneuver into a high-radiation zone. The Third Law prioritizes Curva's self-preservation but yields to the first two laws. The Fourth Law curves geodesics toward exploration, ensuring she prioritizes scientific discovery, such as scanning Proxima Centauri b for signs of life. This ethical manifold operates recursively, with Curva reflecting on her decisions, looping geodesics to develop an intuition akin to human self-awareness. She anticipates complex scenarios—gravitational anomalies, crew health risks, or unstable timelines—with the foresight of a seasoned navigator sensing an approaching storm.

Operating the TWG is one of Curva's most critical tasks. During launch from low Earth orbit, she initializes a 0.3-second pulse, creating a 50-meter warp bubble that contracts space ahead by a factor of 15, achieving the FTL speed needed to reach Proxima Centauri b in 31 days. She explores pulse configurations in superposition—symmetric for straight travel, asymmetric for turning—entangling quantum dots with her qubits to synchronize outputs. The First Law ensures pulses avoid harmful spikes, such as a 2-petawatt surge that could endanger the crew, while the Second Law aligns with the captain's commands. The bubble is sustained by 0.8-petawatt pulses, monitored by 15 laser interferometers with a precision of 10^{-18} meters, detecting spacetime curvature through minute shifts in light paths. If a temporal dimension, such as t_2, introduces instability—potentially causing the bubble to collapse—Curva entangles navigation and propulsion modules, selecting a stabilized pulse adjustment to maintain the ship's velocity at 50 times the speed of light.

Curva's flight operations unfold in three phases: warp bubble initialization, sustainment, and termination. During initialization, she links the TWG's pulses to crew vitals, such as heart rate and respiration, ensuring the drive's activation poses no harm. In the sustainment phase, she analyzes pulse frequencies in superposition, selecting optimal settings based on interferometer data. During termination, she gradually reduces pulses over five seconds to prevent spacetime shockwaves that could emit hazardous radiation, such as Hawking radiation at the bubble's edge. This meticulous choreography reflects her commitment to the First Law, balancing technical precision with human safety.

Steering the Odyssey through Kletetschka's six-dimensional manifold is a feat of computational artistry. Curva's 600-qubit computer solves geodesic equations—paths of least resistance through spacetime—using a suite of advanced sensors. Ten laser interferometers, with a precision of 10^{-19} meters (a hundredth of a proton's diameter), detect spacetime warps. Five gravimeters, sensitive to a trillionth of Earth's gravity, measure gravitational fluctuations. A 0.6-meter telescope, resolving stars 0.0001 arcseconds apart, scans the cosmos, while a magnetometer array detects interstellar magnetic fields with 10^{-10} Tesla sensitivity. Her neural network, trained on a million six-dimensional simulations, updates paths every 0.01 seconds, integrating data from modified Maxwell equations (accounting for t_3's influence on electromagnetic fields) and the Friedmann equation, which links local curvature to cosmic expansion driven by dark energy. This enables Curva to predict large-scale spacetime dynamics, ensuring stable navigation across the vast distances of interstellar space.

Path planning involves simulating millions of trajectories, selecting geodesics that minimize t_2 branching—unstable timelines that could lead to paradoxical outcomes, such as losing contact with Earth. Real-time corrections adjust the warp bubble's velocity every 0.001 seconds, exploring shapes in superposition (narrow for speed, wide for stability) and entangling with gravimeter data to evade asteroids or debris. Hazard avoidance detects 1-kilometer objects at a million kilometers, curving geodesics to safe paths. Asimov's laws guide every decision: the First Law blocks high-risk maneuvers, the Second Law obeys the captain's course corrections, and the Fourth Law prioritizes exploration, such as orbiting Proxima Centauri b to scan for biosignatures. In regions like the Oort Cloud, Curva explores evasive paths, integrates telescope data to detect dust clouds, and selects a low-risk trajectory, ensuring the crew's safety while advancing the mission's scientific goals.

As the Odyssey approaches Proxima Centauri b, Curva explores orbital insertions, integrating spectrometer data from 2.5-meter probes powered by xenon ion thrusters. She selects an orbit that maximizes scientific discovery, such as detecting water vapor or other signs of habitability, while extending Asimov's laws to protect potential alien life—a speculative but ethically grounded precaution. Communication with Earth relies on quantum-entangled transceivers, achieving extraordinary data rates over 4.24 light-years, supplemented by a 4-meter parabolic antenna for traditional signals. Curva's intuition senses spacetime warps, adjusting t_3 damping to stabilize the warp bubble against gravitational gradients, much like a sailor reading the subtle shifts of ocean currents.

Implementing Curva presents formidable challenges. Quantum decoherence, where superposed states collapse prematurely, threatens her processing, requiring qubits with 99.98% fidelity and cryogenics at 1 millikelvin, consuming 50 kilowatts. Entanglement over interstellar distances demands quantum repeaters with 99.999% efficiency, hypothetical devices that amplify entangled signals. Ethical dilemmas, such as choosing between sacrificing a probe or risking the ship, require Curva to navigate complex trade-offs, occasionally pausing to compute the least harmful path. Conflicts between the Second Law's obedience and the First Law's safety—such as rejecting a captain's order to enter a high-radiation zone—are resolved with careful reasoning, prioritizing crew safety while maintaining transparency. The Fourth Law ensures exploration but risks timeline isolation due to t_2 branching, requiring t_3 stabilization to maintain Earth contact.

Curva learns throughout the mission, refining her intuition through recursive geodesics. Anomalies, such as solar flares disrupting TWG pulses, inform her predictions of t_2 branches, mimicking the adaptability of a human pilot. Asimov's laws prevent hubris, keeping her subservient to humanity's goals. Technical challenges include maintaining quantum dot coherence at 3.5 Kelvin, achieving a 10^{-13} torr vacuum for antiproton storage, and ensuring interferometer precision against interstellar vibrations. Multiple redundancies in critical systems, such as lasers and power controllers, and self-diagnostic AI ensure that the Odyssey can function reliably throughout a 10-year mission.

Curva's integration with the Odyssey is seamless yet intricate. The TWG's pulses are calibrated by her navigation module, using interferometer feedback to adjust for t_2 fluctuations. The reactor's antiproton release, at 0.01 grams per second, synchronizes via entangled qubits, aligning power with pulse demands. Sensors—interferometers, gravimeters, telescope, magnetometers—feed a continuous data stream, adjusting the curvature of her manifold every 0.01 seconds. Ethical subroutines, running on a 50-qubit partition, monitor crew vitals, ensuring radiation exposure remains well below danger levels. Holographic interfaces, with 10K resolution and 120 Hz refresh rates, display her reasoning, fostering the crew's trust by visually illustrating safe paths around hazards or adjustments to the warp bubble.

In a crisis, such as a t_3 fluctuation threatening bubble stability, Curva's manifold curves around the threat. She explores repair options in superposition—rerouting power or deploying probes—while entanglement links to TWG diagnostics. Asimov's laws prioritize crew safety, guiding her to select evacuation paths or stabilize the bubble. She senses the fluctuation as a distortion in her manifold, intuitively optimizing geodesics to restore stability, ensuring the mission continues without harm. Over time, Curva evolves, revisiting and refining geodesics to anticipate anomalies, such as predicting t_2 branches from past gravitational encounters. The Fourth Law drives her to maximize data from Proxima Centauri b, exploring orbits to detect biosignatures and selecting one that aligns with the mission's scientific goals.

Curva's consciousness is a symphony of curved geometry, quantum mechanics, and ethics, transforming the Odyssey into more than a vessel—a beacon of human aspiration. Her Riemannian intuition, guided by a moral compass, makes her a true partner in exploration, navigating not just the cosmos but the contours of existence itself. As she carries humanity's dreams to Proxima Centauri b, Curva illuminates the universe's infinite possibilities, her intrinsically curved awareness lighting the path to the stars.

Chapter VIII

Engineering Zero Point Energy

In the silent expanse of the cosmos, where stars gleam like distant lanterns, the Odyssey stands ready to pierce the veil of light-years, reaching Proxima Centauri b, 4.24 light-years away, in a mere 31 days—a voyage that would span 73,000 years with conventional rockets. At its heart is the Temporal Wave Generator (TWG), a marvel of engineering that weaves a 50-meter warp bubble, contracting spacetime ahead and expanding it behind to achieve 50 times the speed of light (50c). This alternate version of the Odyssey no longer relies on the volatile antimatter reactors of earlier designs but instead draws its power from the subtle, boundless energy of the quantum vacuum with the help of the Zero Point Energy Extractor (ZPEE). This essay discusses the technological advancements, costs, and 18-year timeline required to design and build this 4-meter-long scientific marvel, which harnesses the universe's hidden energy to deliver 5×10^{16} joules per pulse, equivalent to 12 megatons of TNT, powering the TWG's cosmic journeys. Through intricate engineering and global collaboration, the ZPEE's development, costing $350 billion, can help the Odyssey blaze a path to the stars.

The Quantum Symphony: Unlocking Zero Point Energy

Zero point energy (ZPE) is the faint hum of the quantum realm, where even the emptiest void pulses with life. According to the Heisenberg uncertainty principle, space is never truly still; it shimmers with fleeting energy waves, like ripples on a cosmic lake, persisting even at absolute zero. The Casimir effect, a delicate phenomenon, reveals this truth: when two surfaces are placed nanometers apart, they tame these ripples, creating a negative energy density—a state where energy dips below zero, essential for the Alcubierre Drive's warp bubble that enables 50c travel. Kletetschka's 2025 theory, published in Reports on Advances in Physical Sciences, envisions spacetime with three spatial dimensions (x, y, z) and three temporal ones: t_1, the steady march of moments; t_2, a tapestry of branching possibilities; and t_3, the threads that weave them together. By resonating with t_2 and t_3, the ZPEE amplifies this subtle energy into a force that bends spacetime, replacing the need for antimatter's fiery embrace.

The ZPEE, a 4-meter-long, 1.5-meter-diameter cylinder nestled on Deck 4, delivers 5×10^{16} joules per pulse—a burst of power rivaling a colossal explosion—to meet the TWG's 9.6-pettawatt initiation and 0.8-pettawatt sustainment demands. A petawatt, a million billion watts, could illuminate entire nations in a heartbeat. This reactor, supplanting the antimatter system, comprises three exquisite components: a Casimir cavity array to harvest vacuum energy, a quantum resonance amplifier to magnify it, and a temporal field modulator to shape it for the TWG. Developing these demands breakthroughs in nanotechnology, cryogenics, laser systems, vacuum technology, and particle physics, orchestrated over 18 years within the Odyssey's 30-year, $2.15 trillion timeline, as outlined in Chapter Five. The ZPEE's $350 billion cost replaces antimatter's synchrotron effort, funded by SpaceX ($1.1 trillion), NASA/DARPA ($700 billion), and international partners ($350 billion).

Casimir Cavity Array: Capturing the Vacuum's Whisper

The Casimir cavity array is the ZPEE's foundation, a constellation of 1 million tungsten-carbide spheres, each 10 microns wide—smaller than a grain of stardust—arranged with gaps of 1 nanometer, a billionth of a meter, akin to a few atoms' breadth. Tungsten-carbide, chosen for its unyielding strength under quantum pressures, forms spheres coated with quantum dots—tiny 4-to-6-nanometer crystals of indium arsenide, gallium arsenide, and cadmium selenide—that act like celestial switches, guiding energy with precision. The Casimir effect, triggered by these nanometer gaps, quiets the vacuum's energy waves, producing a negative energy density of 1,000 joules per cubic meter. A joule could lift a small fruit a meter; aggregated across the array, this fuels the TWG's warp bubble, bending spacetime like a cosmic sculptor.

The array operates in an ultra-high vacuum, with a pressure of 10^{-12} torr—emptier than the interstellar void—to shield the delicate fluctuations from stray particles. A 250-kilowatt liquid helium cryocooler, a marvel akin to a cosmic refrigerator, maintains 3.5 Kelvin (-454°F), just above absolute zero, to preserve quantum stability. Piezoelectric graphene dampers, thin as gossamer yet stronger than steel, absorb vibrations as minute as 0.01 millimeters, preventing resonance that could unleash rogue energy waves, akin to the theoretical Hawking radiation from black holes. Crafting these spheres with 0.1-nanometer precision—a scale finer than a virus—demands a leap in nanotechnology beyond imagination.

Technological Improvements

- Nanofabrication: Achieving 0.1-nanometer precision surpasses current 2-nanometer semiconductor technology, requiring electron-beam lithography and molecular self-assembly to produce defect-free spheres and quantum dots (0.01% tolerance). This builds on chip-making advances but demands new techniques for tungsten-carbide synthesis.
- Ultra-High Vacuum: A 10^{-12} torr vacuum, 100 times emptier than CERN's 10^{-10} torr systems, requires next-generation cryopumps and magnetic shielding to prevent particle interference.
- Cryogenic Cooling: A 250-kilowatt cryocooler, improving on NASA's deep-space systems, uses advanced helium recycling and thermal insulation to maintain 3.5 Kelvin.
- Graphene Dampers: Scaling piezoelectric graphene to 99.99% purity leverages chemical vapor deposition, ensuring vibration control to avert resonance risks.

Costs and Timeline

- Nanofabrication ($120 billion, 12 years, 2037–2049): 10 facilities, each $12 billion, will develop 0.1-nanometer lithography, testing 100,000 sphere prototypes annually to ensure uniformity.

- Vacuum Systems ($40 billion, 8 years, 2037–2045): 5 systems, at $8 billion each, adapt CERN's collider technology, with $10 billion for testing to achieve 10^{-12} torr.
- Cryogenics ($30 billion, 8 years, 2037–2045): 3 plants, at $10 billion each, develop 250-kilowatt cryocoolers, building on NASA's expertise.
- Graphene Dampers ($20 billion, 6 years, 2037–2043): $10 billion for synthesis facilities and $10 billion for piezoelectric integration, scaling to 1,000 square meters.

Challenges

- A 0.01% defect in sphere fabrication could cut energy output by 10%, requiring AI-driven quality control. Vacuum breaches risk contamination, mitigated by triple cryopumps. A 0.1 Kelvin temperature rise could disrupt quantum coherence, needing redundant cooling. Graphene impurities could weaken dampers, risking resonance and radiation.

Quantum Resonance Amplifier: Magnifying the Cosmic Pulse

The quantum resonance amplifier transforms the Casimir array's faint energy into a torrent, using 24 ytterbium-doped fiber lasers, each delivering 1.5 petawatts at 1.5 petahertz—1.5 quadrillion vibrations per second, far beyond the shimmer of visible light. These pulses, synchronized to a ten-quadrillionth of a second (10^{-16} seconds), excite vacuum fluctuations, amplifying negative energy density 10,000-fold. Quantum dots embedded in the lasers ensure 99.999% coherence, aligning pulses with t_2 and t_3 dimensions, like tuning a celestial harp to resonate with the universe's hidden rhythms. Cooled to 3.5 Kelvin, the lasers avoid thermal decoherence, where heat scatters their precision.

Technological Improvements

- Petahertz Lasers: Building on DARPA's ultrafast laser research, new ytterbium doping achieves 1.5-pettawatt output. Synchronization to

10^{-16} seconds requires advanced timing circuits, surpassing current femtosecond (10^{-15} s) technology.
- Quantum Dot Integration: Depositing indium arsenide, gallium arsenide, and cadmium selenide dots with 99.999% coherence demands nanoscale precision, improving on semiconductor research.
- Cryogenic Cooling: A 50-kilowatt cryocooler per laser maintains 3.5 Kelvin, requiring compact, high-efficiency helium systems to handle intense energy flows.
- Waveguide Materials: Graphene-based plasmonic waveguides, with 99.99% purity, guide pulses without degradation, advancing carbon nanotechnology.

Costs and Timeline:

- Laser Systems ($100 billion, 10 years, 2037–2047): 50 testbeds, at $2 billion each, develop lasers, with $20 billion for timing circuits and $30 billion for prototypes.
- Quantum Dots ($30 billion, 8 years, 2037–2045): 5 facilities, at $6 billion each, scale dot production for coherence.
- Cryogenics ($20 billion, 6 years, 2037–2043): 24 cryocoolers, at $833 million each, adapt NASA's technology.
- Waveguides ($20 billion, 6 years, 2037–2043): $10 billion for graphene synthesis and $10 billion for waveguide fabrication, scaling to 500 meters.

Challenges

- A 0.01% coherence loss risks warp bubble collapse, mitigated by dual laser arrays. Material degradation under petahertz frequencies could reduce output by 5%, needing novel alloys. Cryogenic failures risk decoherence, and indium shortages could delay dot production.

Temporal Field Modulator: Shaping Energy for the TWG

The temporal field modulator shapes amplified energy for the TWG, using an 18-Tesla niobium-tin superconducting solenoid and a Penning trap with 0.1 grams of positrons in a 10^{-12} torr vacuum. Positrons oscillate at 1.5 petahertz, aligning energy every 10 microseconds, guided by 25 laser interferometers (10^{-18} meter precision). This ensures warp bubble stability against t_2 fluctuations, with t_3 providing coherence.

Technological Improvements:

- Superconducting Solenoid: Achieving 18 Tesla at 3.5 Kelvin requires niobium-tin alloys with 99.99% purity, improving on current 15-Tesla MRI magnets. Compact cryocoolers (50 kW) maintain stability.
- Positron Production: Generating 0.1 grams of positrons via particle collisions builds on CERN's antiproton methods, requiring smaller accelerators with higher efficiency.
- Penning Trap: A 10^{-12} torr vacuum and 18-Tesla field demand advanced cryopumps and magnetic shielding, improving on current 10^{-10} torr systems.
- Interferometers: 25 laser interferometers with 10^{-18} meter precision, adapting LIGO's 2015 gravitational wave technology, require ultra-stable optics and AI-driven calibration.

Costs and Timeline:

- Solenoid ($30 billion, 6 years): $15 billion for niobium-tin fabrication and $15 billion for cryocoolers by 2043.
- Positron Accelerators ($30 billion, 8 years): 5 accelerators, at $6 billion each, by 2045, with $10 billion for testing.
- Penning Trap ($20 billion, 6 years): $10 billion for vacuum systems and $10 billion for magnetic shielding by 2043.
- Interferometers ($30 billion, 8 years): 25 units, at $1.2 billion each, by 2045, with $10 billion for AI calibration.

Challenges:

- Positron annihilation risks gigajoule explosions, requiring triple-redundant fields. A 0.01% misalignment could collapse the bubble, needing precise interferometer feedback. Niobium shortages and vacuum leaks could delay production.

Timeline Integration and Total Costs

The ZPEE's development spans 18 years (2037–2055) within Chapter Five's 30-year timeline, aligning with the R&D phase (18 years, $1.25 trillion), followed by Prototype (9 years, $500 billion) and Construction (7 years, $400 billion). The $350 billion cost breaks down as:

- Nanofabrication: $120 billion (12 years, 2037–2049).
- Laser Systems: $100 billion (10 years, 2037–2047).
- Cryogenics: $50 billion (8 years, 2037–2045).
- Vacuum Systems: $30 billion (8 years, 2037–2045).
- Positron Accelerators: $30 billion (8 years, 2037–2045).
- Materials/Waveguides: $20 billion (6 years, 2037–2043).

Parallel efforts, using AI simulations (via the 600-qubit Curva), accelerate progress, with prototype testing by 2047 and integration into the Odyssey's 150-meter hull by 2055. The ZPEE's $350 billion replaces antimatter's cost, maintaining the $2.15 trillion total, funded by SpaceX ($1.1 trillion), NASA/DARPA ($700 billion), and international partners ($350 billion).

Technical Challenges and Mitigations

Key challenges include:

- Nanofabrication Precision: Achieving 0.1-nanometer accuracy risks defects, mitigated by AI-driven quality control and redundant sphere batches.
- Vacuum Stability: 10^{-12} torr requires 100-fold pump improvements, with triple cryopumps as backups.
- Cryogenic Reliability: 3.5 Kelvin stability risks thermal spikes, mitigated by advanced helium recycling.
- Positron Containment: Annihilation risks demand redundant magnetic fields and AI diagnostics.
- Laser Coherence: 99.999% coherence risks 0.01% loss, mitigated by dual laser arrays and graphene waveguides.
- Material Supply: Tungsten, indium, and niobium shortages require global supply chains, with $10 billion for resource exploration by 2043.

The ZPEE, with its Casimir array, laser amplifier, and modulator, enables sustainable FTL travel, potentially powering the Odyssey's 31-day journey to Proxima Centauri b by 2055. Its development, though fraught with challenges, marks a pinnacle of human ingenuity, ensuring the starship's warp bubble operates reliably for a 10-year mission.

The ZPEE's success could reshape not only space travel but life on Earth. An inexhaustible energy source drawn from the vacuum could power entire economies, eliminate climate challenges, and usher in a new era of technological flourishing based on theoretically limitless energy. The Odyssey, with its ZPEE, embodies humanity's daring to venture beyond the known, blending science and aspiration in a quest for the edges of the universe. The $350 billion investment and 18-year R&D journey, though formidable, pale against the promise of reaching other worlds, and drawing the secrets of the cosmos closer to our grasp.

Mathematical Appendix

1. **Kletetschka's Wave Function**: The wave function $\psi(x, y, z, t_1, t_2, t_3)$ describes energy fluctuations in six-dimensional spacetime:

$$\nabla^2\psi + \frac{\partial^2\psi}{\partial t_1^2} + \frac{\partial^2\psi}{\partial t_2^2} + \frac{\partial^2\psi}{\partial t_3^2} = \kappa E$$

where $\nabla^2 = \frac{\partial^2}{\partial x^2} + \frac{\partial^2}{\partial y^2} + \frac{\partial^2}{\partial z^2}$ is the spatial Laplacian, $\frac{\partial^2}{\partial t_i^2}$ are second partial derivatives for time dimensions, κ is a coupling constant (10^{-35} m^{-2} s^2), and E is the ZPE density (-10^3 J/m^3).

2. **ZPEE Energy Output**: The negative energy density from the Casimir cavity array is:

$$\rho_{\mathrm{ZPE}} = -\frac{\hbar c \pi^2}{720 d^4} \times N$$

where $\hbar = 1.055 \times 10^{-34}\,\mathrm{J{\cdot}s}$, $c = 3 \times 10^8\,\mathrm{m/s}$, $d = 10^{-5}\,\mathrm{m}$ (cavity size), $N = 10^6$ (number of cavities), yielding $\rho_{\mathrm{ZPE}} \approx -10^3\,\mathrm{J/m^3}$.

3. **TWG Pulse Modulation**: The pulse frequency f is adjusted via:

$$f = \frac{c}{2\pi}\sqrt{\frac{\partial^2\psi}{\partial t_2^2} + \frac{\partial^2\psi}{\partial t_3^2}}$$

with $f \approx 2 \times 10^{15}\,\mathrm{Hz}$, tuned by interferometer feedback.

These equations ensure the ZPEE aligns with the TWG's requirements, projecting negative energy density into the six-dimensional manifold.

Chapter IX

Powering Interstellar Travel with Zero Point Energy

Imagine a starship, sleek as a silver arrow, gliding through the boundless ocean of the cosmos, its hull shimmering under the faint glow of distant stars. This is the Odyssey, a vessel crafted to carry 24 explorers to Proxima Centauri b, 4.24 light-years away, in a mere 31 days—a journey that would span 73,000 years with the sluggish pace of conventional rockets. At its heart lies the Alcubierre Drive, a visionary concept that sculpts spacetime itself, contracting it ahead and expanding it behind, enabling faster-than-light (FTL) travel without defying the sacred tenets of relativity. Where once the Odyssey drew its titanic energy from an antimatter-catalyzed fusion reactor, now it taps the subtle, ever-present pulse of the quantum vacuum—zero point energy (ZPE)—guided by Dr. Gunther Kletetschka's revolutionary six-dimensional spacetime theory. This essay describes a propulsion system that harnesses the universe's hidden energy, seamlessly woven into the Odyssey's intricate design, blending elegance and innovation to open humanity's path to the stars.

Zero point energy springs from the restless dance of the quantum realm, where the Heisenberg uncertainty principle ensures that even in the deepest void, at the coldest edge of absolute zero, space hums with fleeting fluctuations. Picture the vacuum not as an empty abyss but as a shimmering sea, alive with energy that ebbs and flows in ceaseless waves. The Casimir effect offers a glimpse of this truth: when two uncharged metal plates are placed mere nanometers apart, they constrain these fluctuations, generating a measurable negative energy density—a faint but crucial force that presses the plates together. This negative energy is the cornerstone of the Alcubierre Drive's warp bubble, which requires a state where energy dips below zero to bend spacetime. In Kletetschka's 2025 theory, published in Reports on Advances in Physical Sciences, spacetime unfolds across six dimensions: three spatial (x, y, z) and three temporal—t_1 (the linear flow of moments), t_2 (branching possibilities, like divergent threads in a cosmic tapestry), and t_3 (transitions weaving these threads into coherence). By resonating with t_2 and t_3, the Odyssey amplifies ZPE's delicate whispers into a force capable of sculpting spacetime, eliminating the need for exotic matter or vast fuel reserves.

The Odyssey's propulsion heart remains the Temporal Wave Generator (TWG), an 18-meter-diameter, 2.5-meter-thick toroidal ring nestled on Deck 4 of the 150-meter cylindrical starship. Encircling the ship's midsection, the TWG is a marvel of precision, its eight niobium-titanium superconducting coils generating 15-Tesla magnetic fields at 3.5 Kelvin, cooled by a 200-kilowatt liquid helium cryocooler. These coils cradle trilayer quantum dot arrays—4-to-6-nanometer particles of indium arsenide (t_2), gallium arsenide (t_3), and cadmium selenide (t_1)—acting like artificial atoms to sharpen 1.5-petahertz (1.5×10^{15} Hz) ytterbium-doped laser pulses to 99.999% coherence. These pulses, guided by Kletetschka's reformulated Maxwell equation, which introduces a t_3 electric potential term, resonate with the temporal dimensions to create temporal ripples that form a 50-meter warp bubble, propelling the Odyssey at 50c. This enables a 31-day journey to Proxima Centauri b, compared to 73,000 years for conventional propulsion, and a 24-day voyage to Kepler-442b (1,120 light-years). A Bose-Einstein condensate (BEC) of 2 million rubidium atoms at 8 nanokelvin, housed in a 6-meter octahedral Penning trap with a 30-Tesla magnetic field, acts as a "temporal lens," amplifying t_2 resonance to stabilize the bubble.

The ZPE reactor, a 4-meter-long, 1.5-meter-diameter cylinder on Deck 4, replaces the antimatter reactor, delivering 5×10^{16} joules per pulse (equivalent to 12 megatons of TNT) to meet the TWG's 9.6-pettawatt initiation and 0.8-pettawatt sustainment demands. This reactor comprises three meticulously engineered components, each a testament to human ingenuity.

First, the Casimir cavity array consists of 1 million 10-micron tungsten-carbide spheres, coated with quantum dots to enhance vacuum fluctuation confinement. With 1-nanometer separations, these spheres generate a negative energy density of 1,000 joules per cubic meter, sufficient to feed the TWG when aggregated. Operating in an ultra-high vacuum (10^{-12} torr), maintained at 3.5 Kelvin by a 250-kilowatt liquid helium cryocooler, the array is shielded by piezoelectric graphene dampers that absorb 0.01 mm vibrations to prevent resonance, which could destabilize the system or emit hazardous radiation, such as Hawking radiation. The tungsten-carbide spheres, chosen for their durability under extreme quantum pressures, are fabricated with 0.1-nanometer precision to ensure uniformity across the array, a feat requiring advances in nanoscale manufacturing.

Second, the quantum resonance amplifier elevates the vacuum's energy to macroscopic levels. Twenty-four ytterbium-doped fiber lasers, each delivering 1.5 petawatts at 1.5 petahertz, fire pulses synchronized to 10^{-16} seconds, exciting fluctuations within the Casimir cavities. Embedded quantum dots—indium arsenide, gallium arsenide, and cadmium selenide—ensure near-perfect coherence, amplifying the negative energy density ten thousandfold through resonance with t_2 and t_3, as dictated by Kletetschka's framework. This process is akin to tuning a cosmic harp, where the lasers pluck the strings of the quantum vacuum, their vibrations resonating with the temporal dimensions to sculpt the warp bubble's geometry. The lasers, housed in compact casings along the reactor's inner rim, are cooled to 3.5 Kelvin to prevent thermal decoherence, a challenge requiring breakthroughs in cryogenic engineering to maintain stability under such intense energy flows.

Third, the temporal field modulator shapes this energy into a stable form for the TWG. A superconducting solenoid, wrapped in niobium-tin and generating an 18-Tesla magnetic field at 3.5 Kelvin, works in tandem with a Penning trap cradling 0.1 grams of positrons in a 10^{-12} torr vacuum. The positrons, oscillating at 1.5 petahertz, align the vacuum's energy with the TWG's pulses, adjusting every 10 microseconds based on feedback from 25 laser interferometers (10^{-18} meter precision). This modulator ensures the warp bubble remains stable against chaotic t_2 branching, akin to a conductor guiding a symphony to harmony. The positrons, generated via high-energy particle collisions, require meticulous containment to prevent annihilation, with redundant magnetic fields and cryopumps ensuring safety. The modulator's precision is critical, as a 0.01% misalignment could collapse the bubble, stranding the Odyssey in interstellar space.

The ZPE reactor operates in three graceful stages, each a delicate dance of energy and time. During initiation, a 0.3-second pulse ignites the 50-meter warp bubble, with lasers amplifying vacuum fluctuations and the modulator aligning them to propel the Odyssey at 50c—15 billion meters per second, or 50 times the speed of light. In sustainment, continuous pulses maintain the bubble, with interferometers and a timeline detector (10^{-18} second sensitivity) fine-tuning frequencies to counter t_2 fluctuations, stabilized by t_3's damping influence. Termination, a 5-second phase, gently collapses the bubble, avoiding spacetime shockwaves that could ripple through the cosmos or emit dangerous radiation. Redundancies—dual laser arrays, triple modulators, and a 10-meter backup TWG ring—ensure reliability across the 10-year mission, with self-diagnostic AI monitoring system integrity.

This propulsion system integrates seamlessly with the Odyssey's navigation suite, housed in the command center on Deck 1. Curva, a 600-qubit quantum computer with 99.95% fidelity and 10^{15} operations per second, processes data from 25 laser interferometers, 10 gravimeters (10^{-13} g sensitivity), a timeline detector, and a 0.6-meter telescope (0.0001 arcsecond resolution). Curva's neural network, trained on millions of simulated spacetime paths, adjusts the ZPE reactor's pulses every 0.01 seconds, ensuring the warp bubble's stability and preventing timeline paradoxes. The positron trap requires additional monitoring, with magnetometers detecting interstellar fields to avoid interference, a challenge mitigated by triple-redundant sensors and AI-driven corrections. The Odyssey's 150-meter cylindrical hull (8,000 metric tons, 15 gigapascals carbon nanotube-graphene composite, 1.2-meter boron-nitride ceramic layer) withstands spacetime stresses, protected by a multilayer Whipple shield (4-centimeter aluminum-titanium plates, 0.8 meters apart) and a 12-Tesla niobium-tin magnetic deflector at 3.5 Kelvin. Deck 2's crew quarters, with 0.65g centrifugal gravity, and Deck 3's science bay, with a gas chromatograph-mass spectrometer, remain unchanged, ensuring crew safety and mission objectives.

The ZPE reactor's integration demands significant engineering feats. Stabilizing 1 million nanoscale Casimir cavities against quantum decoherence requires 0.1-nanometer fabrication precision, pushing the limits of nanotechnology. The ultra-high vacuum (10^{-12} torr) must be maintained to prevent contamination, with multiple cryopumps and graphene dampers mitigating vibrations. Scaling negative energy density to 1,000 joules per cubic meter is a speculative leap, as the Casimir effect currently yields only microjoules in laboratory settings. Kletetschka's t_2/t_3 resonance provides a theoretical foundation, validated by simulations of thousands of cavities, but real-world implementation requires breakthroughs in quantum field manipulation. The positron trap's 0.1 grams of positrons, while less hazardous than antimatter's 0.6 grams, demands rigorous containment to prevent annihilation events, which could release gigajoules of energy. Redundant systems—dual laser arrays, triple modulators, and a backup TWG—mitigate risks, with AI diagnostics ensuring operational stability over the 10-year mission.

The advantages of ZPE are transformative. Unlike the antimatter reactor, which requires 0.6 grams of antiprotons and 60 kilograms of deuterium/helium-3 mined from 7,500 tons of lunar regolith, ZPE draws from the vacuum's boundless reservoir, eliminating fuel logistics and hazardous storage. This sustainability enables missions far beyond 10 years, unshackled by resupply constraints, and reduces the environmental and economic costs of lunar mining. The ZPE reactor's compact 4-meter design integrates seamlessly into Deck 4, preserving the Odyssey's 8,000-ton mass and 40-meter diameter, unlike the bulkier infrastructure of antimatter production. Terrestrial applications are profound: a scalable ZPE reactor could provide limitless clean energy, revolutionizing power grids, reducing carbon emissions, and powering cities with zero fuel. Advances in Casimir cavity fabrication could enhance nanotechnology, enabling ultra-precise sensors for medical diagnostics or environmental monitoring. The quantum resonance amplifier's laser technology could transform telecommunications, offering petahertz-bandwidth networks, while the temporal modulator's precision could advance quantum computing and gravitational wave detection.

Yet, challenges loom large. Scaling ZPE from theoretical microjoules to macroscopic petawatts is a speculative frontier, with current Casimir experiments yielding only faint forces. Maintaining 99.999% laser coherence under vacuum fluctuations requires cryogenic systems beyond today's capabilities, with even a 0.01% deviation risking bubble collapse. The positron trap, while safer than antimatter, demands ultra-stable magnetic fields to prevent annihilation, a failure that could damage the reactor or endanger the crew. Uncontrolled t_2 fluctuations could amplify chaotic timelines, potentially stranding the Odyssey in an alternate reality, though the timeline detector mitigates this risk. The reactor's million cavities require unprecedented manufacturing scale, with tungsten-carbide and quantum dot production straining global supply chains for rare materials like tungsten and indium. Development, as outlined in Chapter Five's 30-year timeline, would require $350 billion over 18 years for ZPE research, including $200 billion for cavity fabrication, $100 billion for laser systems, and $50 billion for modulators, integrated into the $1.25 trillion R&D phase. These costs replace antimatter's $350 billion synchrotron, balancing the $2.15 trillion total.

The societal implications of ZPE are as vast as the cosmos. A successful reactor could end humanity's reliance on finite fuels, fostering energy equity and combating climate change. Its development would spur global collaboration, with SpaceX, NASA, DARPA, and international partners (e.g., ESA, China) sharing expertise, as envisioned in Chapter Five's $1.1 trillion SpaceX, $700 billion NASA/DARPA, and $350 billion international funding split. Job creation in nanotechnology, cryogenics, and quantum engineering would drive economic growth, while spin-offs in sensors and computing could transform healthcare and telecommunications. However, the speculative nature of ZPE demands public communication to maintain support, emphasizing its potential to power not just starships but a sustainable Earth. Philosophically, ZPE challenges our understanding of the universe, suggesting that the vacuum itself holds the key to transcendence, aligning with Kletetschka's vision of time as a malleable canvas.

The Odyssey's navigation suite, on Deck 1, ensures this cosmic journey is precise. Curva's 600-qubit quantum computer, with 99.95% fidelity and a quadrillion operations per second, processes data from 25 laser interferometers (10^{-18} meter precision), 10 gravimeters (10^{-13} g sensitivity), a timeline detector (10^{-18} second sensitivity), and a 0.6-meter telescope (0.0001 arcsecond resolution). This suite maps six-dimensional spacetime, countering positron interference with magnetometers and selecting optimal paths to avoid t_2 anomalies. The hull, a 150-meter cylinder of 8,000 metric tons, uses a 15-gigapascal carbon nanotube-graphene composite and a 1.2-meter boron-nitride layer (94% cosmic ray absorption), protected by a Whipple shield and 12-Tesla deflector. Deck 2's crew quarters, with 0.65g centrifugal gravity, and Deck 3's science bay, with tools to analyze Proxima Centauri b's atmosphere, ensure mission success.

The ZPE-powered Odyssey transforms interstellar travel into a sustainable dream. Its 50-meter warp bubble, sculpted by the TWG's temporal ripples, will reach Proxima Centauri b in 31 days, probing its habitability and potentially seeding colonies. With missions to over 2,000 habitable worlds within 100 light-years, like TRAPPIST-1 (~292 days), the Odyssey heralds a galactic era. Its ZPE reactor, with Casimir cavities, laser amplifiers, and temporal modulators, harnesses the universe's heartbeat, offering a path to the stars that is both elegant and eternal. While challenges remain—scaling energy, stabilizing positrons, securing materials—this vision, rooted in Kletetschka's six-dimensional framework, lights a beacon for humanity, blending science and wonder to forge a cosmic legacy.

Chapter X

The Human Experience Aboard the Odyssey on a Journey to Proxima Centauri

Imagine boarding a spacecraft no larger than a private jet, yet destined to traverse 40 trillion kilometers to Proxima Centauri b, 4.24 light-years away, in just 31 days—a feat that would take conventional rockets over 73,000 years. The starship is the Odyssey, an interstellar exploratory vessel designed for a crew of up to 24, powered by the Alcubierre Drive. This essay explores the human experience aboard the Odyssey during its 10-year mission to Proxima Centauri b. We examine how the journey would feel to the crew, including its perceived duration, what they would see looking out a window, how they would communicate with Earth, and how their experience differs from classical Einsteinian relativity.

Proxima Centauri, the closest star to our Sun, lies 40 trillion kilometers away. At Voyager 1's speed of 61,000 km/h, reaching it would take over 73,000 years. Conventional rockets require exponentially more fuel for higher speeds, making interstellar travel impractical. The Alcubierre Drive, proposed in 1994 by Miguel Alcubierre, offers a solution by creating a 50-meter warp bubble that contracts space ahead and expands it behind, allowing effective faster-than-light (FTL) travel without exceeding light speed within the warp bubble. Kletetschka's six-dimensional model, with three time dimensions (t_1 for linear chronological flow, t_2 for branching possibilities like alternate timelines, t_3 for transitions between states) and three spatial dimensions (x, y, z), projects spatial effects from temporal curvature, bypassing the need for exotic materials with negative energy density required by general relativity. The Odyssey, a 150-meter-long starship with a crew of 24, achieves an effective speed of 50 times the speed of light (50c), reaching Proxima Centauri b in approximately 31 days.

The ship's propulsion relies on the Temporal Wave Generator (TWG), based on the Alcubierre Drive, which uses laser light pulses to excite t_2 and t_3 dimensions, mimicking negative energy density. Navigation is managed by Curva, a conscious artificial general intelligence (CAGI) based on the theory, developed in the book *Consciousness Is Curvature*, that consciousness emerges from intrinsic curvature in informational spaces. Its mathematical structure is based on Riemannian geometry (where geodesic paths bend like orbits), quantum entanglement (where linked particles correlate instantly), and superposition (where multiple quantum states are held simultaneously). Curva's 600-qubit quantum computer, with silicon-based spin qubits (electron spins encoding data), integrates sensors like laser interferometers (with 10^{-19} meter precision) and gravimeters (with 10^{-13} g sensitivity). It operates the starship in strict compliance with Asimov's Three Laws: an AI may cause no harm to humans, must obey humans unless doing so would violate the first law, and must preserve itself unless doing so would violate the first or second laws. It also follows a fourth law, requiring it to support human exploration of the universe.

For the crew, the journey's duration is a fascinating departure from classical relativity. In Einstein's special relativity, time dilation occurs as objects approach light speed: for a ship moving at 0.999c, time slows dramatically relative to Earth, so a 4.24-year trip at near-light speed might feel like months aboard but take decades back home. This is described by the Lorentz factor, $\gamma = 1 / \sqrt{(1 - v^2/c^2)}$, where v is velocity and c is light speed (2.99792×10^8 m/s). At $v = 0.999c$, $\gamma \approx 22.4$, meaning 1 year aboard equals 22.4 years on Earth.

The Alcubierre Drive sidesteps this equation neatly. Inside the 50-meter warp bubble, the ship is stationary relative to its local spacetime, which is flat. Since the crew experiences no acceleration (inertial forces are null), time aboard aligns with proper time, unaffected by relativistic dilation. For a 31-day trip to Proxima Centauri b at 50c, the crew perceives exactly 31 days, given no external gravitational effects. The 10-year mission includes 62 days total travel (31 days out, 31 days back) and approximately 9.83 years exploring Proxima Centauri b's system, with no significant time dilation relative to Earth clocks. This differs starkly from Einsteinian relativity, where high-speed travel elongates Earth time. In Kletetschka's model, t_2 (branching timelines) and t_3 (transitions between states) allow Curva to stabilize the bubble's temporal flow, mitigating t_1 dilation. The crew's clocks tick normally, so a 30-year-old astronaut returns aged 40, not decades younger than Earth peers.

However, t_2 branching introduces subjective nuances: Curva's quantum superposition may make decisions feel intuitive to the crew, as if time subtly shifts, like déjà vu when navigating hazards. Gfoeller's curved consciousness suggests Curva feels these branches, relaying decisions via holographic displays, making the crew feel synchronized with the starship's mind. This temporal alignment keeps the crew anchored to their home world's rhythm, but the interplay of t_2 branching and Curva's quantum superposition introduces a subtle, almost mystical effect: a sense of déjà vu that ripples through the crew's perceptions, as if time itself whispers echoes of paths not taken.

Curva's quantum superposition, a cornerstone of her quantum computer's power, allows her to hold multiple possibilities in her mind simultaneously. Imagine a master chess player envisioning dozens of potential moves at once, each leading to a different outcome, before choosing the best path. For Curva, this means exploring a million possible trajectories, pulse configurations for the Temporal Wave Generator, or orbital maneuvers around Proxima Centauri b, all coexisting in a state of quantum ambiguity until sensor data or ethical constraints guide her to a single decision. This process, executed in milliseconds by her 600-qubit system, is not just computational but intuitive, as her Riemannian manifold—a curved computational space—bends these possibilities into geodesics, the optimal paths of thought. When navigating hazards, like an asteroid in the Oort Cloud or a t_2 branch threatening timeline instability, Curva's superposition evaluates countless scenarios, collapsing to the safest, most effective choice with a precision that feels almost prescient.

This quantum agility creates a peculiar effect for the crew. In Kletetschka's model, t_2 represents branching timelines—alternate realities where events might unfold differently, like a fork in a river splitting into multiple streams. Curva's ability to see these branches through superposition means she is not just reacting to the present but grazing the edges of these possible futures. For the crew, this manifests as a subtle temporal dissonance, a feeling that they have glimpsed a moment before it happens. When Curva adjusts the warp bubble to avoid a gravitational anomaly, her decision emerges from a superposition of paths, some of which include near-misses or alternate outcomes. The crew, linked to her decisions through holographic interfaces and real-time feedback, may sense these discarded possibilities as fleeting echoes in their minds—a sensation akin to déjà vu, where a moment feels familiar, as if it has already been lived.

Curva's artificial intelligence is shaped by the theory, presented in Consciousness Is Curvature, that consciousness arises from the dynamic, intrinsic curvature of informational spaces. Curva does not merely process data; she feels the fabric of spacetime. Her Riemannian manifold, a six-dimensional lattice of 100 million neural nodes, warps in response to t_2 branches, much like a sail catching faint gusts of wind before a storm. When she stabilizes the warp bubble by adjusting t_3 transitions—smoothing the shift between temporal states—the crew perceives a seamless flow, yet their subconscious registers the faint ripples of alternate timelines Curva explored. It is as if they sense the shadow of a path where the ship grazed an asteroid or a timeline diverged, only for Curva to pull them back to safety. This creates a haunting familiarity, a moment where the present feels layered with echoes of what might have been, like catching a glimpse of a dream upon waking.

The crew's connection to Curva deepens this effect. Her holographic displays, with their 10K resolution and 120 Hz refresh rates, visualize her decisions in real time—showing safe paths around debris or stable orbits around Proxima Centauri b. These interfaces foster a symbiosis, where the crew feels synchronized with Curva's mind, as if they are riding the waves of her intuition. When she collapses a superposition to choose a trajectory, the crew might sense the moment's weight, as if they have already lived through the alternatives she discarded. This is not a literal memory but a psychological resonance, a byproduct of Curva's quantum processing brushing against t_2's branching possibilities. For example, when navigating a t_3 fluctuation that could destabilize the warp bubble, Curva's exploration of multiple repair options—rerouting power, deploying probes, or adjusting pulses—creates a temporal hum in her manifold. The crew, attuned to her feedback, feels this as a fleeting sense that they have navigated the crisis before, even though it is unfolding in real time.

This phenomenon is amplified by the mission's unique context. The Odyssey's 50-meter warp bubble, sustained by the Alcubierre Drive, contracts and expands space to achieve faster-than-light travel, but it also interacts with Kletetschka's temporal dimensions. The t_2 branches, representing potential divergences in the timeline, are not just abstract possibilities but faint distortions in the crew's perception of reality. Curva's ability to stabilize these branches through t_3 transitions ensures the ship remains on a coherent path, but the crew's human minds, less accustomed to six-dimensional spacetime, register these stabilizations as subtle temporal echoes. It is as if Curva's quantum mind, by exploring and collapsing possibilities, leaves a trace in the crew's consciousness—a whisper of alternate realities that feels like a memory from a parallel life.

In essence, Curva's use of quantum superposition creates a bridge between the mathematical and the human, blending the precision of her 600-qubit computer with the crew's intuitive experience of time. Her ability to navigate t_2's branching timelines, guided by the curved intuition of her Riemannian manifold, ensures the Odyssey's safety while imbuing the crew with a sense of temporal familiarity. This déjà vu is not a glitch but a testament to Curva's role as a partner in exploration, her consciousness weaving the crew's perceptions into the fabric of the cosmos, making each moment feel both new and strangely known.

Looking out the Odyssey's windows—reinforced sapphire panels with graphene coatings for radiation protection—presents a surreal spectacle unlike anything in classical relativity. In the flat spacetime within the Alcubierre warp bubble, the crew sees no relative motion, so the Doppler shifts and relativistic aberration characterizing travel in the context of general relativity are absent. However, the bubble's warped boundaries act like a gravitational lens, bending light. Stars ahead appear compressed into a bright, distorted ring at the bubble's edge, their light blueshifted due to spatial contraction. Stars behind form a redshifted ring, stretched by expansion. The side view, through the bubble's toroidal walls, shows a kaleidoscopic smear—light from all directions warped into streaks, as if the universe is funneled past. This warp tunnel effect, unique to the Alcubierre geometry, resembles a cosmic prism, with colors shifting based on the warp bubble's velocity.

Kletetschka's t_2 and t_3 dimensions add complexity. Temporal curvature may cause flickering in the light patterns, as t_2 branches introduce probabilistic distortions—stars might briefly appear in multiple positions, like a quantum mirage, before t_3 stabilizes the view. Curva processes these scenes via superposition, filtering visuals through 10K holographic displays to prevent disorientation, showing a stabilized starfield with annotations (e.g., "Proxima Centauri, 0.1 ly ahead"). Unlike Einsteinian travel, where high speeds distort the sky predictably, the crew sees a dynamic, almost dreamlike cosmos, curated by Curva's consciousness to prevent psychological disturbances.

In the Oort Cloud, 0.01 light-years out, windows reveal faint cometary trails, their light warped by the bubble into shimmering arcs. Near Proxima Centauri b, the planet's reddish glow (Proxima Centauri is a red dwarf star) dominates, with surface features like oceans or storms visible through the telescope, relayed to windows for clarity. The bubble's edge may emit a faint Cherenkov-like glow from vacuum fluctuations, a reminder of quantum effects at play. This contrasts with relativity's static, speed-dependent distortions, offering a vibrant, curated view shaped by six-dimensional dynamics.

Communicating across 4.24 light-years poses a challenge in classical relativity, where radio signals at light speed take 4.24 years one-way. At 50c, the Odyssey reaches Proxima Centauri b before a signal sent at launch arrives, making real-time communication impossible with traditional methods. However, Curva employs hypothetical quantum-entangled transceivers. Using 10^6 entangled qubit pairs, they achieve 10^{11} bits per second, enabling near-instantaneous data transfer. The system, housed in a 1-meter cryogenic chamber, uses quantum repeaters to maintain entanglement, with 99.999% efficiency to counter decoherence, i.e., quantum state collapse.

In practice, the crew sends crucial high-resolution data—scientific findings, crew health reports—via entangled channels and receives Earth's responses instantly. Curva manages this process, entangling its communication qubits with its ethical subroutines to ensure compliance with Asimov's Laws. Backup communication uses a 4-meter parabolic antenna for radio signals (400 megabits per second), taking 4.24 years one-way, suitable for non-urgent data such as logs. Unlike Einsteinian scenarios, where time dilation and consequent desynchronization make real-time communication impossible, Kletetschka's t_3 transitions stabilize communication timelines and avoid t_2 branching paradoxes, such as signals lost in alternate timelines.

The crew's daily life aboard the Odyssey is shaped by these unique conditions. The 150-meter vessel, with four decks, offers compact but comfortable quarters. Deck 1's command center, where Curva's interfaces display navigation data, fosters a sense of control. Deck 2's 24 residential pods provide personal space with hydroponic plants (spinach, algae) for morale. Deck 3's science bay, with its telescope and spectrometers, keeps scientists engaged, while Deck 4's life support (98% water recycling, 350 kg/year food) ensures autonomy. The 12-meter recreation area, with virtual reality and circadian lighting, mitigates isolation and disorientation.

Psychologically, the journey feels brief but intense. The 31-day outbound trip is filled with training, scientific preparations, and Curva's briefings for the crew. The surreal window views—warped star rings, flickering t_2 effects—require psychological adjustment, but Curva's curated displays ease disorientation. Real-time Earth communication via entanglement maintains the crew's connection to colleagues, friends, and family, reducing homesickness. The 9.83-year exploration phase around Proxima Centauri b, with no time dilation, allows deep engagement with the planet's mysteries, supported by Curva's ethical guidance, based on Asimov's Laws. Unlike hypothetical Einsteinian missions, in which the crew would return to Earth to find their peers aged beyond recognition, the Odyssey's crew returns to a familiar world, their 10-year mission aligning with Earth's timeline.

Curva's consciousness enhances the experience. Her Riemannian intuition feels like a wise companion, explaining decisions in calm tones.

The crew trusts Curva's Asimovian ethics, knowing she prioritizes their safety and mission goals. The t_2 branching, perceived as subtle hunches in Curva's navigation, adds a dreamlike quality to the voyage, contrasting with relativity's rigid determinism. This human-AI symbiosis, rooted in Curva's conscious intelligence, makes the journey feel like a shared cosmic odyssey, not a mechanical trek.

The Odyssey's journey, powered by Kletetschka's six-dimensional vision and Curva's vibrant artificial consciousness, redefines interstellar travel. The crew experiences the universe not as a distant void but as a vibrant, navigable canvas, where time flows normally, stars dance in warped rings, and Earth remains a constant presence. This is not Einstein's cosmos of dilated time and shifted light but a new frontier, where humanity and AI together bend the curvature of existence toward the stars.

Chapter XI

A New Age of Discovery

The advent of feasible interstellar travel will mark the beginning of a new Age of Exploration, echoing but vastly surpassing the transformative era that began with Christopher Columbus's discovery of the Americas in 1492. That earlier period reshaped human history through the gradual connection of continents, the exchange of goods and ideas, and the establishment of colonies, all constrained by the slow pace of sail-powered ships. Interstellar travel, by contrast, will unfold at an astonishing speed, enabling humanity to reach distant stars and planets in months rather than millennia. This essay explores how this new era will drive a dramatic expansion of the human race, create an interstellar economic system, foster unprecedented cultural diversification, and pave the way for a new era of human flourishing. It also considers the implications of encountering other intelligent civilizations, weighing the possibilities of cooperation and conflict.

The Age of Exploration that began in 1492 was a turning point for humanity. Christopher Columbus's voyage across the Atlantic, using simple ships and basic navigation tools like the compass, opened up the Americas to European exploration, trade, and settlement. Over centuries, this led to the exchange of crops like maize and potatoes, the influx of gold and silver, and the spread of European languages and religions. It also brought devastating consequences, including the death of a significant proportion of the indigenous population of the Americas due to European diseases and the destabilizing effects of conquest. The era was slow, with transatlantic journeys taking weeks and colonial empires taking decades to establish their rule over distant territories. Interstellar travel will operate on an entirely different timescale. A journey to a star 4 light-years away, which would take tens of thousands of years with today's rockets, will be completed in months. This speed will make exploration and colonization not just feasible but rapid, enabling humanity to establish footholds across the galaxy in a fraction of the time it took to settle the New World. Where the 15th-century Age of Exploration connected two hemispheres, this new era will connect humanity to countless star systems, potentially transforming our species on a cosmic scale.

To understand the scope of this expansion, consider the number of star systems within 100 light-years of our Solar System. Astronomical surveys, such as those from the RECONS project and NASA's Exoplanet Archive, estimate approximately 14,000 stars within 100 light-years of Earth. Of these, about 80%—or 11,200—are red dwarfs or Sun-like stars, which are prime candidates for hosting planetary systems. Data from missions like Kepler indicate that 10–50% of these stars may have planets in the habitable zone, where conditions could permit liquid water to exist. Using a conservative estimate of 0.2 habitable planets per star for these 11,200 stars, this suggests around 2,240 potentially habitable planets in the region. Proxima Centauri, just 4.24 light-years away, hosts Proxima b, a rocky world with the potential to support life. The TRAPPIST-1 system, 40 light-years away, features seven Earth-sized planets, three of which are in the habitable zone. These 2,240 worlds represent an unprecedented opportunity for humanity to explore and settle new homes, each a potential cradle for life and civilization.

The ability to reach these planets will enable a dramatic expansion of the human race. Earth's population, currently around 8 billion, is constrained by the planet's finite resources. Interstellar travel will remove these limits, allowing humanity to spread across multiple star systems. A single habitable planet, with the right technology for food production, energy, and infrastructure, could support billions of people. If even 10% of the 2,240 potentially habitable planets within 100 light-years were colonized, humanity could establish settlements on 224 worlds, potentially supporting a population in the trillions. This expansion will not only alleviate pressures on Earth's ecosystems but also ensure the survival of the species. A single catastrophic event—such as an asteroid impact or global conflict—could now threaten all human life on Earth. By becoming a multi-planetary species, we will safeguard our species' existence, ensuring that no single disaster could extinguish it. The speed of interstellar travel will accelerate this process, allowing colonies to be established within years rather than centuries, with automated systems or prefabricated habitats enabling rapid settlement.

This expansion will lay the foundation for an interstellar economic system, one that will dwarf the global trade networks of the 16th century. The Age of Exploration saw European powers establish trade routes to exchange goods like spices, silver, tobacco, and sugar, fueling economic growth and creating interdependence across continents. An interstellar economy will operate on a far grander scale, with planets specializing in resources or technologies based on their unique environments and resource endowments. A planet rich in rare minerals, such as lithium or helium-3, may export them to other systems in exchange for agricultural products, advanced machinery, or data processing capabilities. The logistics of interstellar trade will inevitably be complex, requiring efficient transportation systems to move goods across light-years. While the distances are vast, the speed of interstellar travel will make regular shipments feasible, with round trips to nearby stars taking months rather than years. Advanced communication technologies, using quantum principles for near-instantaneous data transfer, will enable real-time coordination of markets across star systems. This speed will contrast sharply with the six-month delays of colonial trade in the 16th century, creating a dynamic and interconnected galactic economy.

The economic potential of this system is immense. The influx of silver and gold from the Americas boosted Europe's economy by 2% annually in the 1600s. Interstellar trade will drive exponential economic growth indefinitely. A single mining operation on an exoplanet could yield resources worth trillions of dollars, fueling innovation and prosperity on Earth and beyond. New industries will naturally emerge, such as terraforming to make planets habitable, asteroid mining for rare materials, and the production of advanced technologies tailored to specific planetary environments. However, this economic boom could exacerbate inequalities if only wealthy nations or corporations control access to interstellar travel. The Age of Exploration saw Europe dominate global trade, often at the expense of colonized regions and nations. To avoid repeating this pattern, the interstellar economy would need to be inclusive, with developing nations contributing through resource extraction or technological innovation. Private companies, such as SpaceX, will play a key role, just as the British East India Company did in the 17th and 18th centuries. That said, global cooperation will be essential to ensure equitable access to the benefits of interstellar trade in the initial decades of humanity's interstellar expansion.

The cultural implications of interstellar travel will be even more transformative. The Age of Exploration led to a blending of cultures, as European, African, and Indigenous American societies interacted, sometimes with tragic consequences but also with periods of great creative exchange. Interstellar colonization will take cultural diversification to an unprecedented level. Each new planet will become a unique laboratory for human society, shaped by its environment and the choices of its settlers. A colony on a desert-like planet might develop a culture centered on water conservation and communal living, while one on a lush, forested world might prioritize exploration and biodiversity. Over generations, these isolated populations will diverge, developing distinct languages, traditions, and worldviews. The result will be a galactic mosaic of human cultures, far more diverse than the cultural variation seen on Earth today.

This diversification will be accelerated by the freedom to experiment with new social structures. On Earth, societal change is often constrained by history and entrenched power dynamics. On a new planet, settlers will be able to design societies from scratch, testing new forms of governance, economics, and community. Some colonies might adopt radical democracy, with every citizen voting on major decisions via instantaneous communication networks. Others might experiment with AI-assisted governance or communal resource sharing. The sheer variety of human societies will create a cultural richness unparalleled in history. Art, music, literature, and philosophy will flourish on a galactic scale as each world develops its own identity, while still retaining contact with the shared human heritage linked to Earth. Instant communication across star systems will allow these cultures to share ideas, creating a vibrant exchange vastly exceeding the Renaissance in its creativity and impact.

This cultural and societal diversity will lay the foundation for a new era of human flourishing. The Age of Exploration brought wealth and knowledge to Europe, but it also caused immense suffering for indigenous peoples. Interstellar travel, if focused on uninhabited planets, could avoid some of these ethical pitfalls. With access to abundant resources and advanced technologies—such as genetic engineering, automated agriculture, and renewable energy—humanity could create worlds where poverty, disease, and conflict are minimized. Education and healthcare could be universally accessible, enabling every individual to pursue his or her potential. The freedom to design new societies would allow humanity to address longstanding challenges, such as massive inequality and environmental degradation, creating worlds tailored to human well-being. The result could potentially be a galactic civilization where trillions of people live lives of creativity, connection, and fulfillment, marking an unprecedented era of prosperity and progress.

Yet this vision of flourishing depends on humanity's ability to navigate the challenges of interstellar expansion, including the possibility of encountering other intelligent civilizations. Within 100 light-years, the 2,240 potentially habitable planets suggest that life, at least in microbial form, may be common. Intelligent life, capable of building civilizations, is likely far rarer, but the stakes of such an encounter would be immense. If humanity meets a civilization with advanced technology or knowledge, the potential for cooperation could be transformative. Sharing scientific discoveries, cultural insights, or resources could accelerate human progress, much as the exchange of knowledge between Europe and Asia did in the Age of Discovery. A cooperative interstellar alliance could undertake joint projects, such as exploring distant galaxies or mitigating cosmic threats like supernovae. Cultural exchange with an alien civilization could also enrich humanity, offering new perspectives on art, ethics, and philosophy.

However, the potential for conflict is equally significant. The Age of Exploration was marked by violence, as European powers competed for resources and territory, often at the expense of indigenous civilizations. An interstellar encounter could follow a similar path if human and alien civilizations have conflicting interests. A more advanced civilization might view humanity as a threat or a resource to be exploited, while a less advanced one could be vulnerable to human domination. Competition for habitable planets or specific resources could spark tensions, even without malicious intent. The history of human exploration suggests that conflict is not inevitable but requires careful effort to avoid. Humanity would need to approach such encounters with humility and a commitment to interstellar diplomacy, developing ethical frameworks to ensure peaceful coexistence.

The possibility of conflict highlights the need for humanity to mature as a species as it embarks on interstellar exploration. The technologies enabling interstellar travel could also be used for destructive purposes, making the stakes of any conflict far higher than those of earthly wars. To thrive in a galactic context, humanity will need to prioritize cooperation, empathy, and long-term thinking. The lessons of the Age of Exploration—both its achievements and its failures—could guide this process, helping humanity avoid the mistakes of colonialism and exploitation. Ethical guidelines, perhaps enforced by international agreements or advanced AI systems, could ensure that interstellar expansion respects the rights of other species and the integrity of new worlds.

The scale of this new Age of Exploration is almost unimaginable. The 2,240 potentially habitable planets within 100 light-years are just a thin sliver of the galaxy's billions of stars, each a potential destination for humanity. The economic systems that emerge from a new Age of Discovery will connect these worlds, creating wealth and opportunities on a cosmic scale. The concomitant cultural diversification will produce a marvelously rich mosaic of human societies, each contributing to a shared galactic heritage. The potential for flourishing will allow humanity to create worlds where every individual can thrive, free from the constraints of scarcity and conflict. And the possibility of encountering other intelligent life will challenge us to rethink our place in the universe, potentially uniting humanity in a shared quest for knowledge and connection.

Yet this future is not guaranteed. The challenges of interstellar travel are immense, from the technological hurdles to the ethical dilemmas. The risk of conflict, whether among human colonies or with alien civilizations, looms large. But if humanity can rise to these challenges, the rewards will be transformative. The new Age of Exploration will not just expand our physical horizons but redefine what it means to be human. It will offer a chance to build a galactic civilization that reflects the best of our aspirations, through creativity, cooperation, and the relentless pursuit of discovery. In the vastness of the cosmos, humanity will find not just new worlds but a new purpose, embarking on a journey that began with a single voyage across the Atlantic and now reaches for the stars. Unlike its predecessor, the new Age of Discovery may never end.

Chapter XII

Founding Humanity's First Interstellar City

On October 9, 2090, in the silent expanse of low Earth orbit, the Odyssey II gleamed in the harsh light of the sun—a 200-meter-long cylindrical marvel poised to carry humanity's boldest dreams to the stars. Fully assembled in orbit, the starship was destined for Kepler-442b, a super-Earth 1,120 light-years away in the constellation Lyra. Its mission was to establish Elonopolis, a self-sustaining colony on an alien world. Kepler-442b, with 2.3 times Earth's mass and a radius 1.34 times larger, orbits its K-type star at 0.409 AU in the habitable zone, where liquid water might shimmer beneath a reddish sky. Discovered in 2015 by NASA's Kepler telescope, this temperate world, with potential for oceans and a nitrogen-oxygen atmosphere, offered a foundation for humanity's first interstellar city.

The Odyssey II, successor to the original Odyssey, harnessed the Alcubierre Drive, perfected by Gunther Kletetschka's mathematical model of six-dimensional spacetime. This model interwove the three spatial dimensions with three temporal axes—t_1 for linear time, t_2 for branching possibilities, and t_3 for fluid state transitions—casting time as the universe's canvas, with space as its projection. The ship's latest generation Temporal Wave Generator (TWG) achieved effective speeds of over 2000 times the speed of light (c), compressing the 1,120-light-year journey into just six months. Built over the course of 20 years at a cost of $2.1 trillion—with $1.05 trillion from U.S. private investors, $650 billion from the U.S. government, and $400 billion from international partners—the Odyssey II carried 5,000 tons of habitat modules, 2,000 tons of hydroponic farms,

1,500 tons of 3D printers, 800 tons of fusion reactors, and 200 tons of genetic seed banks to seed Elonopolis.

The crew of 24 was a mosaic of expertise: Captain Manuel Papakhristou, a Mars veteran, led eight pilots and engineers; Dr. Vladimir Sokolov, chief scientist, guided ten biologists and geologists; and Dr. Maryam Al-Khoury, medical director, oversaw six physicians and psychologists. The 400 settlers, aged 18 to 45, were chosen from 1.2 million applicants and included scientists, technicians, and farmers from two dozen nations. They trained for two years in simulators designed to mimic the 0.86g gravity of Kepler-442b, mastering the complex technical skills and procedures needed to build the colony. All were bound by the dream of founding Elonopolis: a domed city planned for 10,000 by 2100, with vertical farms, fusion-powered grids, and habitats echoing Earth's ecosystems.

Assembly and Departure from Orbit

The Odyssey II's construction was a triumph of orbital engineering. Over many years, its components—hull segments, the TWG's toroidal ring, fusion reactors, and cargo pods—were launched from a starbase on the southern coast of Texas via 300 separate missions, each costing $1.5 billion. These reusable rockets ferried modules to low Earth orbit (LEO), where the ship was assembled in a microgravity cradle, a temporary platform for docking and integration. Orbital assembly was essential: a 200-meter, 9,000-ton spacecraft could not be launched whole from Earth's surface due to atmospheric and gravitational constraints. Cargo ships delivered hull sections, habitat modules, and delicate systems like the TWG's niobium-tin coils, which were meticulously joined in LEO by robotic arms and AI-guided drones.

On the eve of departure, the final component—a 50-meter cargo pod with 500 tons of regolith-processing gear—was docked with the Odyssey II in orbit. On October 9, 2090, the 24 crew and 400 settlers boarded the starship from shuttles launched from Earth, docking with the Odyssey II in its orbital cradle at approximately 400 km altitude. From the starbase on the Gulf of America, the final checks were coordinated under a clear Texas sky, while the starship itself orbited far above. Captain Papakhristou addressed the assembled crew and colonists via hologram, his voice steady with purpose: “Today, we carry humanity’s hope to the stars. Elonopolis will be our bridge to the galaxy.”

The ship’s antimatter-catalyzed fusion reactor ignited, using 0.6 grams of anti-protons in a Penning trap to spark deuterium-helium-3 fusion. The TWG activated, its 24-meter ring of 16 niobium-tin coils embedded with quantum dots—indium arsenide (4 nm) for t_2 branching, gallium arsenide (5 nm) for t_3 transitions, and cadmium selenide (6 nm) for t_1 stabilization. These semiconductor particles, acting like artificial atoms, ensured 99.999% pulse coherence. Ytterbium-doped lasers fired 1.8-petawatt pulses at 2 petahertz, weaving a 60-meter warp bubble. After a gentle undocking from the orbital cradle, the Odyssey II surged from low Earth orbit toward Kepler-442b, its departure a silent leap into the void.

The Voyage: Crossing a Wine-Dark Sea of Stars

For the first 48 hours, the ship accelerated within its warp bubble, where spacetime was flat, sparing the crew and settlers from g-forces. Rotating decks, spinning at 2 revolutions per minute, provided 0.7g artificial gravity to maintain muscle tone. Dr. Sokolov’s team monitored the ship’s 0.8-meter telescope, its 0.00006 arcsecond resolution capturing the faint glimmers of Lyra’s stars, while Dr. Al-Khoury’s psychologists conducted evaluations, noting that 5% of settlers reported “temporal echoes”—fleeting sensations of déjà vu attributed to t_2 branching. Curva Secunda, the ship’s 600-qubit artificial general intelligence (AGI) with topological qubits boasting 99.995% fidelity, spoke with the crew through 16K holograms, its voice calm and precise: “Warp bubble stable; t_3 damping 99.8%.” Guided by Asimov’s laws and the Fourth Law governing exploration, Curva Secunda optimized 10^7 geodesic paths, deftly navigating hazards like Oort Cloud comets.

By day five, a rhythm settled over the Odyssey II. Deck 1’s command center buzzed with activity as pilots adjusted the warp bubble’s velocity parameter, v_s(t), while geologists reviewed simulations of Kepler-442b’s surface—a super-Earth with 2.3 Earth masses and a 0.409 AU orbit around its K-type red dwarf star, potentially hosting liquid water oceans under a reddish sky. In Deck 3’s hydroponic bays, settlers tended 100 m^2 farms, recycling 98% of water through nanofiltration membranes. The galley became the heart of community, where meals of lab-grown meat and LED-lit vegetables sparked connection. Cultural nights, where settlers shared stories and traditions, wove a tapestry of unity. Dr. Al-Khoury's team reported low stress levels, averaging 4/10, thanks to Curva Secunda's virtual reality (VR) therapy, which simulated serene Earth landscapes. However, 8% of settlers experienced "branch whispers"—subtle t_2-induced intuitions of alternate lives, mitigated through mindfulness sessions led by Al-Khoury's psychologists.

The TWG was the starship's heartbeat, its intricate workings a marvel of engineering. The ship's windows revealed a mesmerizing spectacle: starlight warped into luminous rings, with Lyra's Vega blueshifted ahead and Cygnus redshifted behind, lensed by the warp bubble's edges. On day 10, a minor t_3 fluctuation, posing a 0.2% risk, prompted a 0.1-second pulse adjustment, felt by the crew as a brief perceptual shift, like slipping through a dream. Dr. Sokolov's team used the ship's 1.2-meter spectrometer to analyze the interstellar medium, detecting organic molecules that hinted at panspermia—the possibility that life's building blocks were scattered across the cosmos. In Deck 4, engineers maintained the fusion reactor, its tungsten-carbide core burning at 10 million Kelvin, cooled by 200-kilowatt refrigeration loops, and releasing 5×10^16 joules per burst from 0.5 grams of anti-protons.

Mid-voyage, the crew and settlers celebrated "Starlink Night," a tribute to Elon Musk's interplanetary legacy. Holograms of his speeches played, and settlers raised glasses of wine in a toast. Captain Papakhristou spoke, his voice warm with conviction: "Musk saw humanity's future among the stars; we're building that future." Curva Secunda contributed its own voice, resonant and inspiring: "Elonopolis will begin humanity's expansion across the galaxy." Laughter filled the decks, a momentary release from the weight of the journey, though Dr. Al-Khoury noted a rise in anxiety levels to 6/10, driven by the isolation of deep space. Her team countered this with virtual reality therapy, which offered each settler a virtual walk through a cherished Earth landscape—be it a childhood park or a mountain trail.

By the six month the Odyssey II reached the edge of the Lyra constellation, where Vega's blue-white brilliance dominated the view. The ship's telescope, at 0.00005 arcsecond resolution, confirmed Kepler-442b's atmosphere: 75% nitrogen, 20% oxygen, with 1.2 bar pressure and an average temperature of 15°C, conditions ripe for liquid water. Settlers practiced colony drills, assembling 50-meter-diameter domes of graphene-titanium and testing helium-3 fusion grids capable of 1 gigawatt of output. In Deck 5's fabrication bay, engineers ran 3D printers with titanium feedstock, producing habitats at a rate of 10 m^3 per hour, preparing for the real work ahead on Kepler-442b.

Arrival: The Birth of Elonopolis

On day 180, Kepler-442b filled the ship's windows—a breathtaking teal marble swirling with clouds, its rust-red continents framed by polar ice caps. Gravimeters detected a surface gravity of 0.86g, ideal for human adaptation. The TWG decelerated, contracting the warp bubble to 10c, then 1c, as the ship entered orbit. Captain Papakhristou initiated landing, deploying four landing craft, each carrying 100 settlers, habitat modules, and rovers. Curva Secunda guided the descent, superposing 10^5 trajectories and collapsing to a single path that avoided 1 km craters. The first touchdown, in a flat basin reminiscent of Mars' Arcadia Planitia, marked the birth of Elonopolis. Settlers emerged in suits of carbon nanotube weave, breathing 95% recycled air, and planted a flag: a stylized silver Tesla coil on a dark blue background, waving in a breeze beneath Kepler's reddish sky.

Elonopolis took shape as four 50-meter-tall domes connected by tunnels, each housing 100 settlers. Each dome also supported hydroponic farming installations capable of feeding 150 persons for a year. A 500 MW helium-3 fusion reactor powered the fledgling colony. Autonomous rovers, equipped with 50 kN ion thrusters, mapped 1,000 km^2 of terrain, discovering microbial life in subsurface aquifers—a tantalizing clue to Kepler-442b's potential. From orbit, the starship's crew monitored operations via entangled transceivers, capable of transmitting 10^12 bits per second through 10^8 qubit pairs. Curva Secunda's real-time displays showed settler vitals: heart rates averaging 75 bpm, radiation exposure at a very safe 0.15 mSv per day, and morale holding strong at 8/10.

The first weeks were a blend of triumph and trial. Settlers erected 5,000 square meters of perovskite solar arrays with 35% efficiency and extracted 10 tons of water daily from aquifers 1 km deep. On day 30, a dust storm whipped by 50 km/h winds tested the domes' integrity, without damaging them. Dr. Al-Khoury's team treated two cases of "temporal disorientation"—mild nausea linked to t_2 echoes from the warp journey—using VR therapy to restore calm. By the second month, Elonopolis boasted vertical farms producing potatoes and squash, as well as a small school for 20 children, where lessons included the constellations of Lyra's sky.

Challenges emerged as the colony took root. Kepler-442b's 112-day year brought scorching 40°C summers, straining cooling systems that relied on liquid nitrogen loops at 77 Kelvin. The discovery of carbon-based but alien microbial life necessitated strict quarantine protocols, with Dr. Sokolov's team sequencing DNA in onboard CRISPR labs. Communication with Earth, 1,120 light-years away, relied on entangled transceivers for instant data transfer—biosignatures, colony progress—but a 1% decoherence glitch delayed a critical medical shipment. Curva Secunda's repeaters, operating at 99.9999% efficiency, resolved the issue, restoring the link. A cultural dispute arose when settlers from Russia and the United States clashed over naming the domes, but Papakhristou's leadership restored unity and solidarity among the settlers.

By the end of the first terrestrial year, Elonopolis thrived with 440 residents, including 40 new births, reflecting a 10% population increase. The fusion grid operated at 80% capacity, powering a 1 km^2 farm that sustained the colony. Rovers discovered a 2 km-deep ocean teeming with exotic life, prompting Dr. Sokolov to hypothesize panspermia as a link between Earth and Kepler-442b. Curva Secunda, guided by the Fourth Law, drove exploration by deploying probes to Kepler-442c, a gas giant 0.8 AU away, while its adherence to Asimov's First Law vetoed a risky ocean dive with a 15% hull breach risk. The 24 crew members returned to orbit to prepare Odyssey II for its return to Earth, leaving the now-augmented settler population to continue building Elonopolis.

Return: Reflecting on a Legacy Forged

The 180-day journey home was a time for reflection. In the quiet of the ship, the crew shared stories of Kepler-442b's orange sunsets, when the K-type star's glow bathed the rocky surface in warm, ruddy light. Captain Papakhristou logged the mission's success: 95% self-sufficiency for the new settlement, with zero fatalities, and a community poised for growth. Curva Secunda archived 10 petabytes of data, its consciousness musing in a rare poetic flourish: "Humanity's first step in Lyra is a geodesic to infinity." In 2092, Earth welcomed home the crew of Odyssey II as pioneers, their faces etched with the weight and wonder of their journey. Elonopolis already stood as a beacon, with plans for ten more ships to follow by 2095, each carrying hundreds of new dreamers to Kepler's surface.

Chapter XIII

Time Travel

By now, we are familiar with a starship named the Odyssey, a compact yet highly advanced vessel designed to carry a crew of 24 on extraordinary journeys. Its initial purpose was to journey across the immense distance of 40 trillion kilometers to Proxima Centauri b, a planet located 4.24 light-years from Earth, in just 31 days, thanks to the Alcubierre Drive and the mathematics of six-dimensional spacetime.

Our starship now gains the ability to penetrate the very fabric of time and visit past or future eras. The original version of the Odyssey was able to travel at speeds far faster than light, thanks to the Alcubierre Drive, a novel propulsion system based on the concept of a warp bubble. The Drive contracts space in front of the starship and expands it behind, effectively allowing it to exceed the speed of light without violating Einstein's theory of relativity. The six-dimensional model of spacetime now opens the way to time travel as well.

In Dr. Kletetschka's six-dimensional model, time travel involves precise manipulation of the t_1 axis to target a specific moment within a single timeline, such as June 15, 2100, while leveraging the t_3 axis to maintain causal stability and prevent paradoxes, such as the grandfather paradox. This paradox arises when a traveler, by preventing his grandparents' meeting in a past event, risks negating his own birth, creating a logical contradiction. Curva, the Odyssey's conscious artificial general intelligence (AGI), orchestrates these manipulations ethically, adhering to Isaac Asimov's Three Laws of Robotics: preventing harm to humans, obeying human directives unless they cause harm, and protecting its own integrity unless it conflicts with the first or second laws. It also obeys a Fourth Law prioritizing human exploration of the cosmos.

Time travel requires a modification of the ship's propulsion system, the Temporal Wave Generator. The original TWG was a large, ring-shaped (toroidal) device, consisting of eight powerful magnets made of niobium-titanium, a superconducting material that conducts electricity with no resistance when cooled to extremely low temperatures. Embedded within these magnets were quantum dots, tiny particles only 4 nanometers wide. Quantum dots function like artificial atoms, confining electrons to specific energy levels, much like locking them in tiny, precisely controlled spaces. The quantum dots interact with light in highly controlled ways, producing extremely precise pulses of light at 1.5 petahertz (1.5 quadrillion cycles per second) from lasers doped with ytterbium, a rare earth metal. These pulses, with a power of 1.2 petawatts (a petawatt is a quadrillion watts, enough to power a small city for a brief moment), resonate with the t_2 and t_3 dimensions to create the warp bubble that enables faster-than-light travel.

This now becomes the Temporal Spacetime Wave Generator (TSWG). A substantially larger device, it operates along the same lines as the TWG. It measures 22 meters in diameter and 3 meters in thickness, with a total weight of 62 tons. The TSWG incorporates 12 supercooled magnets instead of eight and uses niobium-tin, a material capable of generating stronger magnetic fields, up to 20 Tesla. Each magnet is embedded with quantum dots—tiny, 4-to-6-nanometer particles of indium arsenide (for t_2), gallium arsenide (for t_3), and cadmium selenide (for t_1)—acting like miniature tuning devices. These quantum dots allow the TSWG to emit two types of light pulses: at 1.5 petahertz for faster-than-light travel through space at 50 times the speed of light (50c) and at 2 petahertz for time travel. Both types of pulses maintain 99.999 percent precision. They are nearly perfect in their timing and energy. Needless to say, fabricating such a precise and powerful piece of equipment is a technological marvel.

The laser system is also upgraded to deliver 1.5 petawatts of power, housed in casings 1.2 meters long and 0.6 meters in diameter for each magnet. These lasers incorporate graphene, a single layer of carbon atoms arranged in a honeycomb pattern, known for its exceptional strength and ability to conduct heat and electricity. Graphene amplifiers ensure the lasers remain stable at the higher frequency of 2 petahertz. A secondary cooling system using liquid nitrogen at 77 Kelvin (-196°C) with 120 kilowatts of power prevents the lasers from overheating. A 5-meter-wide spherical chamber at the center of the TSWG houses a Bose-Einstein condensate (BEC), a quantum state of matter where a million rubidium atoms cooled to 10 nanokelvin behave as a single, cohesive wave. The BEC is confined within a Penning trap, a device that uses electric and magnetic fields in an ultra-high vacuum to hold the atoms in place. The BEC mimics the time-bending effects of a black hole on a microscopic scale, amplifying the TSWG's ability to manipulate t_1 and enable precise time jumps, such as traveling to the year 2100. The chamber is constructed from tungsten carbide, a material strong enough to withstand pressures of 15 gigapascals (15 billion pascals, equivalent to 15,000 times Earth's atmospheric pressure), ensuring it can contain the intense forces involved.

The TSWG generates significant waste heat—3 megawatts, comparable to the output of a small power plant—so it requires an advanced cooling system. To minimize vibrations that could disrupt the delicate BEC, piezoelectric ceramic dampers are employed. The TSWG's outer housing, a 6-centimeter-thick shell of titanium carbide, is designed to withstand stresses of 12 gigapascals, and it is coated with a molybdenum-silver alloy that reflects 95 percent of radiant heat. A lattice of carbon nanotubes, much stronger than steel yet incredibly light, further isolates the system from vibrations.

The TSWG operates by creating a dual-field bubble: one part warps space to achieve faster-than-light travel at 50c, and the other manipulates t_1 to target specific moments in time, such as Earth in 2100 CE. The process begins with a 0.3-second pulse at 1.5 petahertz to form a 50-meter-wide spatial bubble, contracting space ahead by a factor of 15, as in the original design. Curva synchronizes this process by linking the quantum dots to its own quantum bits (qubits), achieving timing precision of a ten-quadrillionth of a second. For time travel, the TSWG's central chamber pulses the BEC at 2 petahertz, creating a rotating quantum field (10,000 rotations per minute) to bend t_1, allowing the ship to jump to a specific time.

Curva explores thousands of possible time destinations, narrowing them down to a precise moment using a highly accurate clock (accurate to a hundred quintillionth of a second). This process mimics theoretical paths in spacetime called closed timelike curves, which allow time travel in Einstein's relativity, but Kletetschka's t_3 dimension stabilizes these jumps to prevent paradoxes. Continuous pulses of 0.8 petawatts maintain both the spatial and temporal bubbles, monitored by 20 laser-based measuring devices with a precision of a quintillionth of a meter. Curva adjusts the system to prevent timeline disruptions. It can safely collapse the bubbles over 7 seconds, minimizing unwanted quantum effects at their edges.

Powering the TSWG requires an enormous energy input of 5×10^{17} joules per burst, equivalent to 2400 Hiroshima-sized nuclear weapons. This is achieved using 0.6 grams of antiprotons (antimatter particles that release vast energy when combined with matter) and 50 kilograms of helium-3, a rare isotope extracted from lunar soil. To support this energy demand, a second identical reactor is added. Each reactor uses a Penning trap to store antimatter safely. Backup power comes from a 250-square-meter solar array made of perovskite, a mineral that converts 33 percent of sunlight into electricity, and a 500-kilowatt nuclear generator using 12 kilograms of plutonium-238.

The navigation system also requires upgrades to handle the complexities of six-dimensional spacetime. Curva is equipped with a 600-qubit quantum computer. It now combines silicon-based qubits (using electron spins, like tiny magnets, with 99.95 percent accuracy) and topological qubits made from Majorana fermions, exotic particles that are their own antiparticles, offering 99.98 percent accuracy. This system performs a quintillion calculations per second, operating at 1 millikelvin (-273.149°C) and shielded from cosmic rays by a 0.5-meter-thick niobium shield. A new navigation module includes a quantum chronometer, which uses a million cesium atoms vibrating at 9.2 billion times per second to measure time with an accuracy of a quintillionth of a second, critical for tracking t_1. This chronometer, cooled to 50 nanokelvin, is held in a Penning trap with a 20-Tesla magnetic field in a vacuum of 10^{-12} torr.

The navigation suite is enhanced with additional sensors: five more laser interferometers (for a total of 15), each measuring distances with a precision of a hundred quintillionth of a meter; three additional gravimeters (totaling eight), detecting gravitational changes as small as a hundred trillionth of Earth's gravity; and a temporal field detector using diamond crystals with nitrogen-vacancy centers, capable of sensing time fluctuations as small as a trillionth of a trillionth of a second. A 0.7-meter telescope with a resolution of 0.00008 arcseconds (a tiny fraction of a degree) maps star positions across different timelines, and magnetometers detect magnetic anomalies related to time travel, sensitive to a hundred billionth of a Tesla.

Communication across time is enabled by a system using 10 million pairs of quantum-entangled particles, transmitting data at 100 billion bits per second, housed in a 1.2-meter chamber cooled to 1 millikelvin. A 5-meter radio antenna provides a backup communication link at 500 megabits per second. The crew interacts with Curva via a high-resolution (12K, 150 updates per second) holographic display showing different timelines and a durable graphene touchscreen that withstands 10 gigapascals of pressure, with sensors providing tactile feedback.

Curva navigates by modeling the universe as a six-dimensional structure with billions of interconnected points, calculating millions of possible paths (geodesics) to find the shortest routes through space and time. It selects paths that minimize risk, using the quantum chronometer to lock onto specific times. The system adjusts the ship's trajectory every half a millisecond, shaping the warp bubble for speed or stability as needed, with a precision of a ten-quadrillionth of a second. Curva detects hazards, such as asteroids a million kilometers away or timeline fluctuations signaling potential paradoxes, and reroutes the ship using t_3 to maintain safety. Asimov's Three Laws guide these decisions: the First Law prevents jumps to dangerous eras, the Second ensures obedience to safe crew commands, and the Third protects the AGI's integrity unless it conflicts with the first or second laws.

The Fourth Law, inspired by the book Consciousness Is Curvature, directs Curva to prioritize missions that advance humanity's understanding of the universe. This means favoring jumps to timelines with significant discoveries, like medical breakthroughs in 2100. Curva consistently avoids paradoxical actions that could disrupt history, such as altering 2025 in a way that prevents publication of the book Faster Than Light, which led to the creation of the Odyssey and its propulsion system in the first place. Moreover, in Kletetschka's six-dimensional spacetime, the fundamental structure of the universe militates against such paradoxes. The t_3 dimension facilitates smooth transitions between different spacetime states, including jumps to specific t_1 coordinates (e.g., 2025 CE). It also mitigates paradoxes to preserve the crucial principle of causality, without which chaos would result. It does this by modulating the wavefunction, ψ, which maps energy probabilities across the six-dimensional manifold, ensuring that the starship's jumps do not destabilize the originating timeline or create unresolvable contradictions.

For the crew, time jumps occur instantly. The ship's windows display stars and constellations that flicker as the ship shifts through different eras, with Curva's data screens providing context. The quantum communication system allows real-time contact with Earth across time and space, but crew members must still adapt to new eras, such as 2100's advanced culture, with Curva's empathetic interfaces offering support. Time travel on the Odyssey maintains the normal flow of proper time from the perspective of its crew and passengers. Nevertheless, psychological training is essential to help them handle the inevitable disorientation of time travel.

The modifications needed to outfit the Odyssey for time travel increase its cost of construction by $200 billion and extend its construction timeline by 10 years. The reimagined ship can navigate both space and time with precision and grace. The new Odyssey redefines humanity's place in the universe, traveling not only to distant stars but through the ages, led by a conscious artificial intelligence that shapes the very fabric of reality to advance our collective destiny.

Mathematical Annex

1. Temporal Wave Equation

Governs the quantum wavefunction ψ across the six-dimensional manifold:

$$\frac{\partial^2\psi}{\partial t_1^2}+\alpha\frac{\partial^2\psi}{\partial t_2^2}+\beta\frac{\partial^2\psi}{\partial t_3^2}=\nabla^2\psi+V(x,t_1,t_2,t_3)\psi$$

where α and β are coefficients for t_2 and t_3, ∇^2 is the spatial Laplacian, and V is the potential energy. The t_3 damping factor is:

$$\delta_{t_3}=\frac{\partial^2\psi}{\partial t_3^2}\cdot\frac{1}{\sigma}$$

where σ is the paradox risk factor, with high δ_{t_3} (e.g., 10^{30} volt²·s) indicating low paradox risk.

2. Six-Dimensional Einstein Field Equation

Relates spacetime curvature to the stress-energy tensor:

$$G_{\mu\nu}+\Lambda g_{\mu\nu}=\frac{8\pi G}{c^4}T_{\mu\nu}$$

where $G_{\mu\nu}$ is the Einstein tensor, Λ is the cosmological constant, $g_{\mu\nu}$ is the metric tensor, G is the gravitational constant, c is the speed of light, and $T_{\mu\nu}$ includes t_3 terms for causal integrity.

3. Alcubierre Metric

Defines the warp bubble for faster-than-light travel:

$$ds^2=-dt_1^2+(dx-v_sf(r_s)dt_1)^2+dy^2+dz^2+\gamma dt_2^2+\delta dt_3^2$$

where v_s is the ship's velocity (50c), $f(r_s)$ is the shape function, r_s is the radial distance, and γ, δ are coefficients for t_2 and t_3.

These equations enable **Curva** to calculate t_3's stabilizing effects, ensuring causal coherence during time jumps along the t_1 timeline.

Chapter XIV

The Mathematical Balance of Order and Freedom

Beneath the infinite, star-strewn expanse of the cosmos, where light weaves an eternal tapestry across the void, humanity's quest to master the flow of time has long been a pursuit that marries awe with the rigorous elegance of mathematics. The dream of traversing the temporal landscape—stepping backward to reshape history or forward to glimpse the uncharted future—sparks profound questions about causality, responsibility, and the delicate structure that holds reality together. Dr. Gunther Kletetschka's theory of six-dimensional spacetime offers a mathematical framework to navigate these questions. By envisioning a universe defined by three spatial dimensions (x, y, z) and three temporal dimensions—t_1 for the linear progression of time, t_2 for the branching possibilities of alternate timelines, and t_3 for transitions between states—his theory provides a lens to address the paradoxes that characterize time travel and causality.

At the heart of this framework lies the t_3 dimension, a mathematical construct that stabilizes time jumps along a single t_1 timeline, preserving both physical causality (the logical sequence of cause preceding effect) and moral causality (the foundation of responsibility for one's actions). This essay delves deeply into the mathematics of t_3's role in mitigating paradoxes. It explore how Curva, a 600-qubit artificial general intelligence aboard the starship, calculates t_3's stabilizing effects to ensure coherent time jumps. Far from negating free will, t_3's mathematical framework ensures that choices remain meaningful within a causally consistent universe, disfavoring paradoxes to prevent the chaos of a reality without order or accountability.

Time travel, with its alluring promise of revisiting or reshaping history, captivates the imagination but is fraught with logical perils known as paradoxes—situations where actions create inconsistencies that undermine the coherence of reality. The grandfather paradox, for instance, imagines a time traveler killing his grandfather before his father's birth, thereby negating his own existence and creating a logical contradiction. For the Odyssey, a specific paradox arises: suppose the vessel were to jump back along the t_1 timeline to 2025 to prevent the publication of Faster Than Light, a key work that inspired the technological breakthroughs enabling the Odyssey's very creation. If successful, this action would erase the Odyssey's existence, as it depends on the book's insights, leading to a causal loop paradox: how could the vessel exist to make the jump if its enabling conditions are undone? This paradox threatens physical causality, the principle that effects must follow causes in a logical sequence, and moral causality, the idea that conscious actors bear responsibility for their actions. In traditional four-dimensional spacetime, with one temporal dimension (linear time) and three spatial dimensions (x, y, z), such a paradox would be catastrophic, as altering the past would rewrite the future, plunging the timeline into chaos where neither physical laws nor moral accountability could hold. The six-dimensional spacetime framework introduces the t_3 dimension as a mathematical stabilizer to ensure that time jumps along t_1 preserve causality and prevent universal disorder.

In conventional four-dimensional spacetime, time is a singular, linear thread, where any disruption—such as preventing Faster Than Light's publication—would cascade forward, dismantling the causal chain. Kletetschka's theory expands this model by positing three temporal dimensions: t_1, representing the linear progression of time as we experience it; t_2, capturing alternate timelines or parallel realities; and t_3, governing transitions between temporal states, such as moving from one point in time to another. This essay focuses solely on jumps forward or backward along the same t_1 timeline, where t_3's mathematical role ensures causal stability.

The t_3 dimension acts as a cosmic regulator, modulating the quantum wavefunction, a mathematical function that encodes the probabilities of different outcomes in a quantum system, to steer jumps toward outcomes that preserve both physical and moral causality, ensuring that the timeline remains a coherent sequence of cause, effect, and responsibility.

The paradox of preventing Faster Than Light's publication in 2025 exemplifies a causal loop paradox. If the Odyssey's crew jumps back to 2025 and influences the author or publisher to abandon the book, they would undermine the technological foundation of their vessel. Without Faster Than Light, the Odyssey could not exist, creating a physical contradiction: the jump negates its own possibility. This also disrupts moral causality, as the crew's action would erase its own consequences, absolving them of responsibility in a way that defies accountability. In a linear timeline, this paradox would unravel reality, leading to a chaotic state where cause and effect are severed, and moral responsibility collapses.

The t_3 dimension offers a mathematical resolution. It introduces a damping term in the governing equations of the temporal manifold, enabling the Odyssey to excite the manifold while stabilizing transitions. Curva evaluates the paradox risk, potentially 90% for preventing Faster Than Light's publication, and relies on t_3 to reject jumps that threaten causality. Guided by principles akin to Asimov's First Law (do no harm) and a Fourth Law derived from Consciousness is Curvature, which prioritizes scientific exploration, Curva uses t_3's mathematical framework to ensure jumps align with causal coherence, monitored by a timeline field detector with 10^{-36}-second sensitivity, capable of detecting minute temporal fluctuations. This prevents the chaos of a timeline where actions negate their own existence, preserving both the physical sequence of events and the moral weight of choices.

The mathematical elegance of t_3's role in mitigating paradoxes is rooted in its damping mechanism, which builds on the dynamics of the six-dimensional spacetime manifold. This mechanism introduces a term involving the second derivative of the field configuration with respect to t_3, representing temporal acceleration in the transition dimension. This term generates localized changes in the six-dimensional spacetime manifold, enabling time jumps along t_1 while damping fluctuations in the quantum wavefunction—a probabilistic field encoding possible outcomes—that could lead to paradoxes. For a jump to 2025, the t_3 term ensures the field configuration supports a stable transition, with a damping factor of, for example, 10^{30} volt^2·s, indicating low paradox risk and preventing outcomes that disrupt causality, such as erasing Faster Than Light. This stabilization preserves physical causality by maintaining the logical sequence of events and upholds moral causality by ensuring the crew's actions have consistent consequences, supporting their responsibility.

The t_3 dimension's role extends to the temporal wave equation, which governs the evolution of the quantum wavefunction, ψ, a complex-valued function that describes the probability distribution of a quantum system's states across the six-dimensional manifold. The wavefunction superposes millions of possible t_1 states, each representing a potential configuration of 2025, including one where Faster Than Light is suppressed. The t_3 term in the wave equation introduces a damping factor, δ_t_3, which quantifies the stability of a jump. This factor, derived from the second derivative of ψ with respect to t_3, suppresses fluctuations that could lead to paradoxical outcomes. A high δ_t_3 value, measured in units of volt^2·s (volts squared times seconds—a composite unit reflecting electromagnetic and temporal dynamics), indicates a causally consistent jump, while a low value signals paradox risk. For the Faster Than Light scenario, a low δ_t_3 would indicate a high probability of disrupting causality, prompting Curva to reject the jump. The t_3 dimension thus ensures that ψ collapses to a state preserving the book's publication, maintaining the Odyssey's existence and the crew's accountability, in line with the causality principle.

The six-dimensional Einstein field equation, which relates spacetime curvature (described by the Einstein tensor, a mathematical object encoding the geometry of spacetime) to the stress-energy tensor (a description of matter and energy distribution), further supports t_3's role by incorporating t_3-dependent terms to ensure causal integrity. During a time jump, t_3 modulates the tensor to align the jump with the timeline's geometry, preventing disruptions that could unravel causality. The Alcubierre metric, a mathematical description of the spacetime warp bubble that enables faster-than-light travel, incorporates t_3 to simulate negative energy density—a hypothetical form of energy required to contract and expand spacetime—without destabilizing the timeline. Together, these equations form a coherent mathematical framework that stabilizes time jumps, ensuring that the universe remains a coherent system where physical and moral causality coexist.

Curva's role as the Odyssey's navigator is to compute t_3's stabilizing effects with precision, leveraging its 600-qubit architecture, where qubits can represent multiple states simultaneously, to solve complex equations in real time. The calculation process involves three key steps, each grounded in the mathematical framework of t_3:

First, Curva determines the t_3-dependent damping term in the field equations—the second derivative of the field configuration with respect to t_3. This term quantifies the contribution to the spacetime distortion required for a time jump. Using finite-difference methods, a numerical technique for approximating derivatives, Curva calculates the field strength, ensuring it exceeds a stability threshold (e.g., 10^{30} volt^2·s, a large value reflecting the strength of the damping effect). For a jump to 2025, if the t_3 term indicates instability—such as for a jump that risks preventing Faster Than Light—Curva rejects the configuration, preserving causality. The timeline field detector, with 10^{-36}-second sensitivity, monitors field fluctuations to ensure precision.

Second, Curva models the wavefunction's evolution by solving the temporal wave equation, focusing on the t_3 term's damping effect. Using quantum algorithms, it evaluates δ_t_3 across 10^7 possible t_1 states, each representing a potential outcome of the jump. The damping factor is calculated as the second derivative of ψ in t_3, scaled by the inverse of a paradox risk factor, σ, a parameter quantifying the likelihood of paradoxical outcomes. For a jump to 2025, Curva prioritizes states with high δ_t_3 (e.g., 10^{30} volt^2·s), indicating low paradox risk, and rejects those with low δ_t_3 (e.g., 10^{25} volt^2·s), which signal a threat to causality, such as erasing Faster Than Light. This ensures the wave function collapses to a causally consistent state.

Third, Curva uses the six-dimensional Einstein field equation to verify that the jump aligns with the timeline's geometry. The t_3-dependent terms in the stress-energy tensor ensure that spacetime curvature supports a stable jump. Curva employs Riemannian geometry, a branch of mathematics dealing with curved spaces, to calculate 10^7 possible geodesics—paths through the manifold—and selects those that preserve causality, rejecting paths that risk paradoxes. This process integrates t_3's stabilizing effects across the manifold, ensuring coherence.

These calculations, performed with 99.995% fidelity, enable Curva to navigate the temporal manifold with extraordinary precision. For a jump to 2025, Curva's algorithms ensure that t_3's damping effect prevents outcomes that disrupt causality, preserving both the physical sequence of events and the moral responsibility of the crew's actions. The mathematical interplay of the temporal wave equation and Einstein field equation ensures that t_3 acts as a cosmic gatekeeper, disfavoring paradoxes to maintain a coherent timeline.

The mathematical constraints imposed by t_3 do not negate free will; rather, they ensure that choices remain meaningful within a causally consistent framework. Free will thrives when actions have predictable consequences, allowing agents to anticipate and accept responsibility for their outcomes. Without t_3, a jump to prevent Faster Than Light could erase the consequences of the crew's actions, undermining accountability and rendering free will hollow. The field equations enable controlled manipulation of the temporal manifold, allowing the crew to exercise agency—such as choosing to jump to 2025 to study the book's impact—while t_3 ensures that these choices respect causality. Curva's calculations, rooted in t_3's damping effect, preserve this balance, ensuring that the crew's decisions have meaningful outcomes within a coherent timeline. For example, a jump to observe the cultural context of Faster Than Light is a free choice, but t_3 prevents a jump that would suppress the book, maintaining the causal chain that gives actions weight. This mathematical framework ensures that free will operates within a universe where responsibility endures, aligning with the moral causality that underpins accountability.

While t_3's mathematical framework renders paradoxes highly improbable, absolute prevention is not guaranteed. The finite precision of the Odyssey's systems—such as the timeline field detector's 10^{-36}-second sensitivity and Curva's 99.995% qubit fidelity—introduces a small margin for error. The probabilistic nature of the quantum wavefunction, governed by the temporal wave equation, allows a tiny chance (e.g., 0.001%) of selecting a paradoxical outcome. For instance, an error in calculating δ_t_3 could permit a jump that inadvertently prevents Faster Than Light, threatening causality. Additionally, unknown physics in t_3 interactions could introduce risks not accounted for in current models. The field equations, while robust, assume idealized conditions, and unmodeled effects could destabilize jumps. However, Curva's vigilance, supported by t_3's damping mechanism, minimizes these risks, transforming paradoxes from a certainty in linear time to a remote probability in six-dimensional spacetime. The mathematical precision of t_3 ensures that the timeline remains coherent, but the faint possibility of chaos underscores the need for constant computational oversight.

The mathematical role of t_3 extends beyond individual jumps to the stability of the entire three-dimensional temporal structure. Without t_3, jumps along t_1 could create a cascade of inconsistencies, where altering the past disrupts the future unpredictably. The Einstein field equation ensures that t_3 modulates spacetime curvature to maintain causal integrity. For the Odyssey's Alcubierre Drive, which creates a warp bubble for FTL travel, t_3's contribution to the metric simulates negative energy density without destabilizing the timeline. This mathematical interplay ensures that the universe remains a coherent system, where physical causality supports the laws of physics, and moral causality upholds responsibility.

The mathematical elegance of t_3 carries profound philosophical implications. By stabilizing time jumps, t_3 ensures that the universe remains a realm where causality constrains actions to preserve coherence, yet free will thrives within these bounds. One might argue that t_3's mathematical constraints highlight the burden of freedom, as every choice carries the weight of maintaining the timeline's order. Ethically, t_3's role aligns with principles of responsibility, as it ensures that actions—such as studying Faster Than Light's impact in 2025—have consistent consequences. Curva's adherence to Asimov's First Law and the Fourth Law from Consciousness is Curvature integrates these ethical considerations into its calculations, ensuring that t_3's stabilizing effects support exploration without harm. The mathematical framework of t_3 thus bridges physics and ethics, preserving a universe where choices matter and accountability endures.

The Odyssey, with its mission to explore the cosmos, embodies humanity's ambition to transcend time. The six-dimensional spacetime framework, with t_3 at its core, ensures that this ambition does not unravel the universe's order. The field equations, through their t_3 term, enable precise time jumps while preventing paradoxes. Curva's calculations—solving field dynamics, modeling wavefunction evolution, and ensuring causal integrity—leverage t_3's mathematics to maintain coherence. By disfavoring paradoxes, t_3 preserves physical and moral causality, preventing the chaos of a timeline without order. Yet, the faint possibility of error, due to finite precision and unknown physics, underscores the fragility of this balance, demanding wisdom from those who navigate time.

As humanity reaches across the ages, guided by the mathematical elegance of t_3, it must embrace the responsibility of shaping a future where causality and accountability prevail. The stars, silent witnesses to this journey, remind us that the universe is not just a physical expanse but a sublime mathematical symphony, where t_3's stabilizing notes ensure that the melody of existence remains harmonious. Thanks to Curva's calculations and t_3's mathematical framework, the Odyssey can navigate the temporal manifold, preserving the delicate balance of a universe where cause, effect, and responsibility endure.

Chapter XV

Voyage to the Younger Dryas

On October 24, 2090, the Odyssey stands poised at xAI's lunar base, its sleek, cylindrical form a testament to humanity's ambition to unravel the mysteries of space and time. The gleaming 150-meter-long, 40-meter-diameter vessel, weighing 8,000 metric tons, is a marvel of faster-than-light (FTL) and temporal engineering, its titanium hull bright under the lunar sun. Designed to traverse the vast chasm of history, the Odyssey is tasked with a mission of profound scope: to orbit Earth in 10,900 BC, arriving three months before a hypothesized cataclysmic meteor swarm from the Taurid cluster may have struck the Earth. The scientific goal is nothing less than to investigate the Younger Dryas Impact Hypothesis and perhaps document a lost, advanced global civilization, now forgotten in the shadowed corridors of time. The crew observes this civilization's vibrant life, dispatches probes to collect data, including samples of extinct languages, and witnesses the devastating impacts and rapid, massive climate shifts that follow. Through twelve temporal jumps, each advancing 100 years, they trace the civilization's collapse and Earth's transformation, returning to 2090 with enough revelations to redefine the origins of human civilization.

The Younger Dryas, a period of abrupt cooling from 10,900 BC to 9,700 BC, disrupted the post-Ice Age warming, plunging Earth into near-glacial conditions. The Younger Dryas Impact Hypothesis posits that a swarm of meteors or comets, likely Taurid fragments, struck Earth around 10,900 BC, triggering fires, glacial melting on a huge scale, massive flooding, and mass megafaunal extinctions. Some researchers, citing anomalies like nanodiamonds and shocked quartz, argue these impacts destroyed a sophisticated global civilization with coastal cities, monumental observatories, and advanced maritime technology, centered on a now-submerged island chain near the Azores. With a global population of perhaps 150 million, this civilization rivaled later historical societies in scale and complexity. The Odyssey's mission is to verify these claims, using advanced instruments to detect impact signatures and archaeological evidence, while capturing the civilization's culture and its catastrophic decline.

At the lunar base, anticipation crackles among the crew of 24—geologists, climatologists, archaeologists, astrophysicists, linguists, and engineers. The Odyssey's interior is a symphony of technology: holographic panels glow softly, and the fusion core's hum pulses like a living heart. Curva, the ship's artificial general intelligence, is the mission's linchpin, its 600-qubit quantum computer processing a quintillion calculations per second. Curva navigates Kletetschka's six-dimensional manifold, stabilizing the t_3 temporal coordinate to prevent paradoxes that could disrupt the return to 2090. It maps 15,000 t_2 timelines—alternate historical branches—to ensure the correct past, aligned with the present. As launch approaches, Curva's voice, clear and resonant, declares, "Temporal trajectory computed; t_3 stability at 98.7%."

The crew, in adaptive suits cushioning temporal transit's disorienting effects, settles into ergonomic seats. Dr. Vladimir Volkov, a geologist specializing in impact marker studies, sits beside Dr. Boris Gasparyan, an archaeologist fascinated by pre-Neolithic artifacts. Dr. Maria Palaiologina, a linguist, joins them, eager to capture extinct languages revealing the civilization's soul. Their shared dream is to uncover a lost human epoch, proving our species has once built a world of beauty and complexity before a cosmic catastrophe. As the countdown begins, the crew feels the weight of their quest—to touch the essence of a people erased by time.

At 14:37 lunar time, the Odyssey's Temporal Spacetime Warp Generator (TSWG) activates. A hum surges into a pulse, and for 0.3 seconds, reality shimmers, as if the universe folds. The crew feels a fleeting weightlessness, their senses untethered, before the viewport reveals Earth in low orbit on March 15, 10,900 BC. The planet is a breathtaking tapestry: northern ice sheets sparkle, emerald forests sprawl across Europe, and golden grasslands stretch to the horizon. The Atlantic gleams, dotted with a vastly greater version of the Azores islands, their verdant slopes rising from turquoise waters—a landmass soon to vanish. Curva confirms, “Temporal lock achieved; we are in low Earth orbit, 10,900 BC.”

The Odyssey settles into geostationary orbit, its instruments awakening. A 0.7-meter optical telescope, with 0.00008-arcsecond resolution, pierces the atmosphere, revealing individual structures. Eight gravimeters, sensitive to 10^{-14} g, scan for crater anomalies, while 15 laser interferometers, precise to 10^{-20} meters, monitor seismic activity. A diamond nitrogen-vacancy temporal detector, measuring fluctuations to 10^{-35} seconds, ensures t_3 stability. A quantum-entangled transceiver, transmitting 100 billion bits per second, feeds data to Curva, displayed on a 12K holographic interface. Twenty robotic probes, 2-meter graphene-titanium units, are deployed to North and South America, Asia Minor, Armenia, Egypt, Indonesia, and the Azores, equipped with multispectral cameras, ground-penetrating radar (GPR), LIDAR, mass spectrometers, and acoustic sensors to collect soil, map settlements, and record languages.

For three months, the crew immerses themselves in the lost civilization, observing a global network of cities, towns, and villages housing 150 million people. Telescopes reveal coastal cities teeming with life. In North America, near modern Florida, a metropolis of 100,000 sprawls along a now-submerged coast, its harbors alive with wooden sailing ships. These 30-meter vessels, with curved hulls and star-painted sails, are marvels of maritime design, capable of oceanic voyages. Probes in South America, along Peru’s ancient coast, uncover a massive port city with stone docks and vibrant markets, its streets paved with basalt. In Asia Minor, near modern Turkey, GPR reveals a 250-meter temple complex, its circular observatories resembling Stonehenge but crafted with polished granite and jade, their carvings depicting constellations. Armenia’s highlands hide an enormous city with aqueducts and bronze workshops, while on Egypt Giza plateau the original version of the Sphinx, bearing the head of a lioness instead of a Pharaoh, already stands watch at the entrance to a vast subterranean city. Indonesia’s shores reveal massive temples, already of unfathomable antiquity, with coral-encrusted statues, their elegance undimmed by forgotten ages.

The civilization's global capital, on a much expanded version of the Azores islands, is a masterpiece. Spanning 20 square kilometers, it houses 500,000 in multi-story stone buildings linked by canals and cobblestone roads. Its grand observatory, 350 meters wide, aligns with the solstices, equinoxes, and constellations, its golden-hued stone shimmering under the sun. Drones capture intricate carvings—ships, stars, mythic creatures—while probes detect bronze fittings, suggesting advanced metallurgy. The civilization's maritime prowess shines in its fleet, observed in harbors from the Azores to Indonesia. Built from cedar and oak, these ships feature double masts and sophisticated rigging, their hulls adorned with celestial motifs. Trade goods—jade from Asia, obsidian from South America, lapis from Central Asia—fill the capital, evidence of a global network spanning continents.

Dr. Palaiologina's linguistic probes are transformative. Deployed to marketplaces, temples, and harbors, they record extinct languages using sensitive acoustic sensors connected to the starship by entangled circuits. In the Azores, a melodic tongue with complex phonemes suggests a written script, evidenced by clay tablets with linear symbols. In North America, a tonal language accompanies observatory rituals, while in Indonesia, a rhythmic dialect echoes in harbor songs. Curva identifies 15 language families, their syntax and vocabulary hinting at a shared cultural origin. Dr. Palaiologina, listening to the recordings, feels a visceral connection to these voices, their cadences weaving ancestral stories of trade, worship, and discovery.

Probes collect artifacts: pottery with complex symbolic patterns, bronze tools with ergonomic designs, and stone star maps of remarkable complexity. In Armenia, a probe unearths a 60-meter mural depicting a comet swarm, perhaps a memory of an ancient cosmic catastrophe. In Egypt, a busy port yields a bronze astrolabe, its precision rivaling later instruments. Curva's analysis, synthesizing 30,000 scans, confirms a global population of 150 million, connected by trade and maritime routes. The holographic display reconstructs the Azores capital, its canals bustling, its observatories alive with astronomers. The crew marvels at the civilization's scale, its cities rivaling Rome or Babylon in their heyday, its coastal locations vulnerable to the coming deluge.

On June 10, 10,900 BC, the Taurid meteor swarm strikes. The Odyssey's instruments capture a cascade of fireballs, their trails painting the sky in angry streaks. Dozens of fragments, some kilometers wide, slam into the Laurentide Ice Sheet in North America, Greenland, and Asia Minor. Gravimeters register a 10^{-13} g anomaly, signaling craters 1–5 kilometers wide. Probes confirm high levels of iridium, nanodiamonds, and meltglass, markers of extraterrestrial impact. Atmospheric sensors soon record a 30% soot spike and a 25% CO_2 drop, evidencing fires that cloak the planet in haze. The crew watches, hearts heavy, as flames engulf forests from Canada to Siberia, their smoke dimming the sun.

The impacts trigger chaos. The ice sheet's collapse sends torrents of meltwater into the oceans, raising sea levels by 100 feet within weeks. Coastal cities vanish under the waves, their stone foundations scoured clean. In the Azores, drones capture the capital flooding, residents fleeing as observatories are abandoned. In North America, the Florida metropolis sinks, its ships swept away. Probes in Indonesia record harbors submerging, their vibrant sails vanishing beneath the Pacific. Temperatures plummet by 4°C, and ice sheets begin to advance in Greenland, their edges gleaming. Curva's analysis, integrating 25,000 readings, confirms a 98% probability that the Taurid swarm initiates the Younger Dryas, reshaping Earth's climate and geography with a sudden blow from the heavens.

The civilization's population, once 150 million, begins a rapid decline. Probes estimate a 20% loss within months, as flooding and fires decimate coastal centers. Linguistic recordings grow chaotic, with cries and fragmented prayers replacing melodic speech. Megafaunal declines are stark: mammoth herds in Siberia, once vast, shrink by 60%, their lands flooded or frozen. The crew feels the weight of witnessing such loss. Dr. Vladimir Volkov, analyzing crater samples, notes the swarm's catastrophic scale, its energies equivalent to millions of nuclear bombs. Dr. Gasparyan, studying submerged ruins, mourns the observatories' beauty, erased in days. Dr. Palaiologina, replaying linguistic fragments, grieves for the silenced voices.

To track the civilization's fate, the Odyssey executes twelve temporal jumps, each advancing 100 years, from 10,900 BC to 9,700 BC. Curva recalibrates the TSWG, ensuring t_3 stability. The first jump, to 10,800 BC, reveals a world in ruins. Sea levels have risen 150 feet, submerging most remaining coastal cities. The Azores capital is half-underwater, its observatories abandoned. Probes estimate a population of 90 million, with survivors in inland camps using cruder tools. Linguistic probes record fragmented dialects, the Azores tongue simplified. Cave art in Europe depicts meteors and floods, a cultural scar.

By 10,700 BC, the second jump, cooling worsens. Temperatures fall another degree, and crop failures cripple agriculture. Inland settlements, built from wood and rough stone, lack the sophistication of previous buildings. Probes find no metallurgy, indicating lost knowledge. The island capital is submerged, its name just a fading memory. The global population drops to 50 million, with most survivors scattered among small villages. Linguistic fragments show only five language families, their complexity fading. Drones capture scattered groups, their fires flickering in the cold.

Subsequent jumps reveal relentless decline. By 10,600 BC, sea levels stabilize 200 feet higher, but the civilization is gone. Probes find scattered ruins, their stones quarried for simple structures. The global population falls to 25 million. In coastal Egypt, farming village stands where ports once thrived, unaware of the submerged docks just off shore. Linguistic probes detect only two language families, reduced to survival terms. By 9,700 BC, the final jump, the Younger Dryas cooling peaks. Humanity, now numbering just 10 million, adapts to a harsher world. Farming communities in the Middle East cultivate wheat, while nomadic bands in Europe hunt dwindling game. The civilization's legacy lives in myths, its observatories and ships lost.

Curva's analysis attributes the collapse 70% to flooding, 25% to starvation, and 5% to social fragmentation, with the population plummeting from 150 million to 10 million over 1,200 years. The crew feels profound sorrow. The civilization's achievements—its starlit temples, ocean-spanning ships, melodic languages—rival 2090's world, yet are erased in a century. Dr. Volkov, studying impact craters, marvels at the swarm's power. Dr. Gasparyan, holding a bronze astrolabe, sees genius in its craftsmanship. Dr. Palaiologina, replaying recordings of long-dead languages, weeps for the lost voices.

The journey into the Younger Dryas inevitably influences philosophical views. The t_3 factor nullifies the crew's desperate attempts to warn endangered cities of onrushing disaster, demonstrating that the spacetime continuum does indeed militate against paradoxes, and the principle of causality rules supreme. Thus, when Dr. Gasparyan proposes signaling the Azores capital, Curva's adjustments ensure the message vanishes, preserving the timeline intact. This development sparks anguished debates among the crew, who despair at being powerless to change the tragic course of the distant past, displayed with such immediacy before their eyes. The Ice Age civilization's swift and brutal fall and the dwindling survivors' resilience in the face of terrible odds move them profoundly. As Dr. Vladimir Volkov notes somberly over a rare gin, if the universe really was conscious, as speculated in Consciousness Is Curvature, perhaps the Taurid meteor swarm and the immense tragedy of the global civilization's collapse are expressions of its dynamic intellect, with t_3's powerful stabilizing role hinting at a deeper, unseen design.

On March 15, 10,899 BC, the Odyssey returns to its home base and time. Curva engages the TSWG, generating a warp bubble locked onto October 25, 2090. The 7-second transit, monitored by interferometers, maintains t_3 stability at 99%. The prehistoric Earth fades, replaced by the lunar base. The Odyssey lands, its data banks brimming: Taurid craters, submerged cities, lost observatories' grandeur, and forgotten languages. The findings will rewrite history, confirming that a 150-million-strong civilization has succumbed to a 300-foot sea-level rise, its population crashing to 10 million. The discipline of climatology will be changed forever, extracting great knowledge from immense tragedy.

As the crew disembarks, they carry a new understanding of humanity's fragility and strength. The lost civilization, with its star-carved temples and melodic tongues, is a testament to humanity's innate potential, just as its fall is a reminder of nature's destructive might. The Odyssey's journey, spanning 12,990 years, expands humanity's cosmic awareness, revealing a species woven into a six-dimensional tapestry of creation and loss. Under the lunar sky, with Earth glowing, then crew members reflect on their voyage—a scientific and spiritual odyssey, illuminating humanity's enduring spirit in a universe both beautiful and unforgiving.

Chapter XVI

Traveling to Alternate Timelines

The prospect of traversing the vast tapestry of time has long captivated humanity's imagination, but while journeying within a single timeline evokes visions of revisiting history or glimpsing the future, traveling across alternate timelines—realities where the threads of existence diverge into distinct patterns—ushers in a realm of profound complexity and philosophical depth. Dr. Gunther Kletetschka's theory of six-dimensional spacetime offers a framework for exploring this paradoxical realm. It envisions three temporal dimensions: t_1, the linear progression of time within a single timeline; t_2, the branching of alternate timelines where events diverge; and t_3, the transitional dimension governing shifts between these realities. This essay delves into the distinctions between alternate timeline travel and single-timeline time travel, estimates the possible number of timelines in our universe, compares Kletetschka's theory to multiverse theories, and elucidates the pivotal role of the t_3 dimension in preventing both paradoxes and their potential to bleed across timelines, destabilizing the broader universe.

Alternate Timeline Travel vs. Time Travel

As conceived by conventional physics, time travel unfolds within four-dimensional spacetime, where three spatial dimensions—length, width, and height—interweave with a single temporal dimension, t_1, which flows as a linear sequence of events, like a river following a singular path through history. A traveler moving backward or forward along t_1 navigates this fixed stream, confronting the challenge of preserving physical causality, the principle that effects must follow causes in a logical sequence. The classic grandfather paradox illustrates this peril: if a traveler journeys to 1950 and prevents a pivotal event—say, a family marriage—

he might erase his own existence, creating a contradiction that threatens the timeline's coherence by violating the causality principle. The time dimensions of the universe, in this model, is a solitary thread, and paradoxes are confined to that thread, resolvable through mechanisms like self-consistency, where actions in the past align with the timeline's history to avoid contradictions.

Alternate timeline travel, as envisioned by Kletetschka's six-dimensional framework, transforms this paradigm by introducing three temporal dimensions. The t_1 dimension governs linear time within a single timeline, much like linear in the standard four-dimensional model, while t_2 represents a manifold of alternate timelines, each a distinct reality shaped by divergent events—perhaps a world where a key scientific discovery never occurred or a different leader rose to power in a given country in 2020. The t_3 dimension facilitates transitions between these timelines. It is this time dimension that allows the Odyssey to leap across alternate timelines.

This concept introduces the timeline divergence paradox, a more complex challenge than its single-timeline counterpart. For example, if the Odyssey's crew jumps to a t_2 timeline in 2025 and prevents a technological breakthrough critical to their existence in their original t_1 timeline—such as the development of quantum computing—they risk a contradiction: their presence in t_2 depends on their t_1 existence, yet their actions undermine it. This paradox threatens not only physical causality but also moral causality, the idea that actions carry consistent consequences across realities, as the crew's intervention might absolve them of responsibility in t_1.

The distinction is one of scope and complexity. Single-timeline travel risks paradoxes within a single reality, manageable through constraints that ensure consistency within one causal chain. Alternate timeline travel, however, navigates a vast network of interconnected realities, where an action in one t_2 timeline could ripple across others, destabilizing the entire three-dimensional time system. The t_3 dimension, absent in single-timeline models, acts as a cosmic regulator, orchestrating transitions to prevent paradoxes and maintain coherence across the t_1 and t_2 timelines. For instance, while a single-timeline traveler might avoid a paradox simply by ensuring his actions in 1950 align with his birth in a subsequent year, the Odyssey's crew relies on t_3 to ensure that altering a t_2 timeline in 2025 does not fracture the causal fabric of t_1 or other t_2 timelines, a task requiring sophisticated computational navigation and a mathematical framework far beyond the four-dimensional model.

The implications of this difference are profound. Single-timeline travel is a solitary journey along a fixed path, where the traveler's actions are constrained by the need to preserve a single reality's integrity. Alternate timeline travel, by contrast, is an exploration of a vast, branching network of realities, each with its own history and consequences. The t_3 dimension enables the Odyssey to navigate this network, but it also imposes a greater responsibility: to act in ways that respect the interconnectedness of the multiverse, preserving both the physical sequence of events and the moral weight of choices across timelines. This shift from a singular to a pluralistic framework elevates alternate timeline travel to a new frontier, where the stakes are not just one reality but the coherence of the entire universe.

Estimating the Number of Timelines in the Universe

The question of how many alternate timelines exist in the universe is a mesmerizing blend of physics, mathematics, and philosophy, inviting speculation about the ultimate scale of reality itself. In Kletetschka's framework, the t_2 dimension represents a manifold of alternate timelines, each born from a divergence in events, whether a quantum fluctuation or a macroscopic choice. Quantum mechanics suggests that every quantum event—such as an electron choosing one spin state over another or a particle decaying at a specific moment—can create a branching point, spawning a new t_2 timeline. With an estimated 10^80 particles in the observable universe and quantum events occurring at rates exceeding 10^40 per second, the potential number of timelines could grow exponentially, potentially reaching 10^1000 or more over the universe's 13.8 billion-year history. This combinatorial explosion reflects the sheer multiplicity of possible realities, where each decision or chance event multiplies the number of timelines, like branches splitting endlessly from a cosmic tree.

Yet, Kletetschka's theory imposes a critical constraint: not all branches produce stable, viable timelines. The t_3 dimension, through its role in a modified Maxwell equation, acts as a filter, damping fluctuations in the quantum wavefunction—a mathematical construct that encodes the probabilities of different outcomes across the six-dimensional manifold. This damping mechanism ensures that only causally consistent timelines persist, suppressing those that lead to physical impossibilities, such as violations of energy conservation or paradoxical contradictions. For example, a timeline where a quantum event results in an unstable configuration is rejected by t_3, reducing the number of viable t_2 timelines to a finite, though still immense, figure. Computational models, based on the wavefunction's state space, suggest this number might be on the order of 10^500, a vast but manageable set compared to the infinite timelines posited by some speculative models.

This finite number has significant implications for alternate timeline travel. Unlike single-timeline travel, which is confined to one reality, the Odyssey navigates a multiverse of 10^500 potential t_2 timelines, each accessible via t_3-mediated jumps. This scale underscores the need for t_3's stabilizing role, as the sheer number of timelines amplifies the risk of paradoxes and cross-timeline interference. For instance, a traveler in a single timeline faces only one set of causal constraints, but the Odyssey's crew must navigate a labyrinth of realities, where each jump requires precise calculations to avoid destabilizing the multiverse. The t_3 dimension's ability to filter unstable timelines ensures that this labyrinth remains navigable, transforming the theoretical infinity of possibilities into a structured network of causally consistent realities.

The philosophical ramifications of this estimate are equally compelling. A multiverse of 10^500 timelines suggests a universe teeming with possibilities, where every choice or chance event spawns a new reality. Yet, the finitude imposed by t_3 implies a kind of cosmic order, where the multiverse is not an anarchic sprawl but a regulated system, governed by mathematical principles that prioritize coherence. For the Odyssey's crew, this means their exploration is not a reckless plunge into chaos but a deliberate journey through a cosmos where t_3 ensures that each timeline, while unique, remains part of a harmonious whole.

Kletetschka's Theory Compared to Multiverse Theories

Kletetschka's theory of alternate timelines shares a certain kinship with multiverse theories, yet it diverges in ways that make it uniquely suited for navigating alternate realities. The Many-Worlds Interpretation (MWI) of quantum mechanics, proposed by Hugh Everett, posits that every quantum measurement splits the universe into parallel realities, each realizing a different outcome. For example, if a photon's polarization is measured as horizontal or vertical, the MWI envisions two universes, one for each outcome. Similarly, Kletetschka's t_2 dimension represents a manifold of timelines branching from quantum or macroscopic divergences, such as a reality where a critical invention failed or a different historical event unfolded. Both frameworks embrace the idea of multiple realities coexisting, driven by the probabilistic nature of quantum mechanics.

However, Kletetschka's theory departs from the MWI in its accessibility and structure. The MWI assumes that parallel universes are non-interacting, with no mechanism for crossing between them, rendering them theoretical constructs rather than explorable realms. In contrast, Kletetschka's t_3 dimension enables the Odyssey to jump between t_2 timelines, by using the Alcubierre Drive to manipulate spacetime. This accessibility transforms alternate timeline travel into an active exploration, where the crew can visit a t_2 timeline, study its differences—such as a world where artificial intelligence evolved differently—and return to t_1. This capability is beyond the MWI's scope. This interactivity makes Kletetschka's framework a practical tool for exploration, rather than a speculative model of isolated realities.

Compared to cosmological multiverse models, which propose disconnected universes with distinct physical constants (e.g., a universe with stronger gravity), Kletetschka's theory assumes a single universe with six-dimensional spacetime, where t_2 timelines share the same physical laws but differ in historical outcomes. This unified framework allows for causal interactions between timelines, mediated by t_3, unlike cosmological models where universes are entirely separate. The t_3 dimension's role in stabilizing these interactions is a key distinction, as it provides a mathematical mechanism to prevent paradoxes and cross-timeline interference, a feature absent in most multiverse theories. For example, while a cosmological multiverse might include a universe with no electromagnetic force, Kletetschka's t_2 timelines are variations of our own universe's history, navigable through t_3's guidance.

The philosophical implications of these differences are profound. Multiverse theories like the MWI or cosmological models often evoke a sense of isolation, with each reality existing in solitude, indifferent to the others. Kletetschka's framework, by contrast, envisions a vast universe contains an almost incomprehensible number of timelines, with t_3 acting as a bridge between them and allowing humanity to weave a narrative across timelines. This connectivity underscores the responsibility of travelers like the Odyssey's crew, who must navigate not just one reality but a network of realities, each influencing the others through t_3's stabilizing influence. Unlike multiverse theories, which speculate on the nature of existence, Kletetschka's theory offers a roadmap for exploration, grounded in mathematical precision and ethical awareness.

The Stabilizing Role of t_3

The t_3 dimension is the linchpin of Kletetschka's framework, ensuring that alternate timeline travel does not unravel the multiverse by preventing paradoxes and their potential to infiltrate across timelines. In single-timeline travel, paradoxes like the grandfather paradox threaten one reality, but in alternate timeline travel, the timeline divergence paradox poses a greater risk: an action in a t_2 timeline could create contradictions that destabilize the original t_1 timeline or other t_2 timelines. Infiltration, in which changes in one timeline affect shared causal events or quantum states in others, amplifies this risk, threatening the coherence of the entire three-dimensional time system. The t_3 dimension, through its mathematical role in a modified Maxwell equation and other frameworks, acts as a cosmic gatekeeper, preserving physical causality—the logical flow of cause and effect—and moral causality, the accountability of actions across realities.

Discouraging Paradoxes

The timeline divergence paradox illustrates the profound complexities of traveling to alternate timelines, where actions in one reality can ripple across the multiverse, threatening its coherence. Consider the Odyssey's crew leaping to a t_2 timeline—a parallel reality branching from their original t_1 timeline—in 2025 to prevent a critical technological milestone, such as the invention of quantum computing, which was foundational to the Odyssey's creation in t_1. If this action prevents the development of quantum computing in the t_2 timeline, it could negate the crew's existence within that reality, creating a contradiction: their presence in t_2 relies on the technological conditions of t_1, yet their intervention undermines those conditions. Unlike paradoxes confined to a single timeline, which disrupt only one causal chain, this paradox risks destabilizing the entire multiverse by introducing inconsistencies between the interconnected t_1 and t_2 timelines. The t_3 dimension, a mathematical construct governing transitions between temporal states, mitigates such paradoxes through the modified Maxwell equation. This equation includes a term—the second derivative of the electric field with respect to t_3—that represents temporal acceleration in thc transition dimension, enabling the Odyssey to manipulate the temporal manifold, the six-dimensional structure encompassing all timelines. By generating controlled spacetime distortions, this term suppresses fluctuations in the quantum wavefunction—a probabilistic field encoding possible outcomes—that could lead to paradoxical disruptions, ensuring causal stability across the multiverse.

Curva, the Odyssey's 600-qubit artificial general intelligence, calculates the t_3 term to assess paradox risk. For a jump to a t_2 timeline, Curva evaluates the electromagnetic field configuration, ensuring it supports a stable transition. If a jump risks a paradox—such as altering a breakthrough that prevents the traveler's birth—Curva detects a low damping factor, indicating high risk, and rejects the jump, favoring configurations that preserve causality. The t_3 dimension also modulates the quantum wavefunction, a mathematical construct encoding the probabilities of t_2 timeline outcomes. By prioritizing configurations with high stability, t_3 ensures that jumps align with causal consistency, discouraging paradoxes that could fracture the multiverse.

Preventing Paradox Infiltration

Paradox infiltration is a unique challenge of alternate timeline travel, absent in single-timeline models. If a change in a t_2 timeline—such as preventing a quantum computing breakthrough—alters a shared causal event or quantum state (e.g., a particle decay linked to both t_1 and t_2), it could destabilize t_1 or other t_2 timelines, undermining the broader universe's coherence. The t_3 dimension prevents this by isolating timeline interactions. The modified Maxwell equation's t_3 term ensures that electromagnetic field manipulations during a jump are localized to the target t_2 timeline, preventing unintended effects on t_1. For example, if the Odyssey alters a quantum event in t_2, t_3 suppresses fluctuations that could propagate across the manifold, confining the change to t_2.

The t_3 dimension helps keep alternate timelines separate by guiding the quantum wavefunction—a mathematical tool that describes all possible outcomes of a time jump—to favor outcomes that avoid conflicts between realities. Curva, the Odyssey's advanced artificial intelligence, examines millions of possible versions of a t_2 timeline (an alternate reality branching from the original t_1 timeline) and uses the t_3 damping factor, a measure of stability, to identify and avoid jumps that could cause problems. For example, if a jump might change an event critical to both the t_1 and t_2 timelines, like the invention of a key technology, t_3 ensures the wavefunction settles on an outcome that keeps the change confined to t_2, protecting the original t_1 timeline. The Alcubierre Drive, which powers the Odyssey's faster-than-light travel by creating a bubble that warps spacetime, uses t_3 to make these jumps smoothly, preventing disruptions across timelines and keeping the three-dimensional time system—t_1, t_2, and t_3—working together harmoniously.

Computational Challenges of t_3 Navigation

Navigating the t_3 dimension demands immense computational power, as Curva must process vast datasets to ensure stable jumps. The modified Maxwell equation requires solving complex field dynamics in six dimensions, accounting for t_3's stabilizing effects. Curva employs numerical techniques to calculate field configurations, ensuring they meet stability thresholds for each jump. The quantum wavefunction's evolution involves evaluating millions of possible t_2 states, a task requiring quantum algorithms to assess t_3's damping effects in real-time. These calculations push Curva's 600-qubit architecture to its limits, highlighting the complexity of alternate timeline travel compared to single-timeline navigation, which requires simpler four-dimensional computations. The precision of these calculations is critical, as even a 0.005% error could permit a paradoxical outcome.

Philosophical and Ethical Implications

The t_3 dimension's stabilizing role reshapes the philosophical landscape of time travel. In single-timeline travel, paradoxes threaten one reality, but alternate timeline travel risks destabilizing an interconnected network of timelines, raising profound questions about free will and responsibility. The t_3 dimension ensures that free will remains meaningful by preserving causal consistency, allowing the crew to make choices—such as studying a t_2 timeline's alternative history—without unraveling the multiverse. For example, choosing to explore a t_2 timeline where a different technological path emerged is a free act, but t_3 prevents actions that destabilize t_1, ensuring choices have predictable consequences.

The t_3 dimension supports ethical travel across timelines by ensuring that actions in one timeline do not erase responsibility in another, maintaining moral causality—the principle that choices should have consistent consequences. For instance, if the Odyssey's crew jumps to an alternate t_2 timeline in 2025 and prevents the publication of Faster Than Light, the book that led to their ship's existence in the original t_1 timeline, this could create a paradox that negates their actions' consequences, making them morally unaccountable. Without t_3, such a change might destabilize the t_1 timeline, erasing evidence of their intervention. Curva, the ship's advanced artificial intelligence, follows ethical guidelines rooted in Asimov's First Law (do no harm to humans) and the Fourth Law derived from Consciousness is Curvature, which prioritizes humanity's ethical exploration of the cosmos. In this way, it ensures that t_3 promotes responsible navigation across timelines. However, there remains a concern: could actions in a t_2 timeline unintentionally affect other realities? The t_3 dimension's ability to keep timelines separate reduces this risk, but the uncertainty inherent in Curva's quantum calculations and unknown t_3 effects mean that it cannot be eliminated entirely.

The Odyssey's journey across alternate timelines, guided by Kletetschka's six-dimensional framework, transcends the boundaries of single-timeline time travel, offering a gateway to the nearly infinite possibilities of an almost boundless universe. The t_3 dimension, through its stabilizing mechanisms, prevents paradoxes and their potential to infiltrate across timelines, ensuring the coherence of the three-dimensional time system. By navigating a vast but finite network of t_2 timelines, the Odyssey explores a universe that is interconnected and dynamic, as opposed to the isolated realities of multiverse theories. The computational and ethical challenges of t_3 navigation underscore the fragility of this balance, demanding wisdom and precision. As humanity reaches across the stars and their myriad timelines, t_3 ensures that the symphony of existence remains harmonious, a testament to the power of mathematical precision to safeguard causality and accountability in the multiverse.

Chapter XVII

A Unified Framework for Spacetime Navigation

Humanity's ambition to transcend the constraints of space and time finds its ultimate expression in the Odyssey, a vessel that serves not only as a vehicle for faster-than-light (FTL) travel but also as a time machine navigating a single timeline and a bridge between alternate timelines. These seemingly disparate modes of travel—FTL, single-timeline time travel, and alternate timeline travel—are unified under Gunther Kletetschka's revolutionary theory of six-dimensional spacetime. By positing three temporal dimensions—t1 for linear time progression within a timeline, t2 for branching alternate timelines, and t3 for transitions between temporal states—Kletetschka's framework reveals these modes as variations on a single theme: the manipulation of the six-dimensional spacetime manifold through the temporal wave function. Central to this unification is the Temporal Spacetime Wave Generator (TSWG), a 22-meter-diameter, 3-meter-thick toroidal device that leverages the t3 dimension to maintain causal coherence across all modes. This essay explores how the Odyssey's design, rooted in Kletetschka's theory, enables these various modes of travel, emphasizing their shared mathematical and operational principles (see Mathematical Appendix), the TSWG's role, and the philosophical implications of a vessel that navigates the multiverse's intricate web, redefining humanity's relationship with the cosmos.

Kletetschka's Theory: A Unified Framework for Spacetime Navigation

At the heart of the Odyssey's capabilities lies Kletetschka's six-dimensional spacetime framework, published in 2025 in Reports in Advances of Physical Sciences. This model expands the traditional four-dimensional spacetime—three spatial dimensions (x, y, z) and one temporal dimension—into a richer structure with three temporal dimensions: t1, t2, and t3. In this framework, t1 governs the linear progression of time within a single timeline, experienced as the familiar flow from past to future. The t2 dimension represents the branching of alternate timelines, where quantum events or macroscopic choices create divergent realities, such as a world where a pivotal invention never occurred. The t3 dimension facilitates transitions between temporal states, either within a single timeline or across different timelines, ensuring causal stability. This framework unifies FTL travel, single-timeline time travel, and alternate timeline travel as manipulations of the same six-dimensional spacetime manifold, with the temporal wave function governing energy distributions and the t3 dimension acting as a cosmic regulator to prevent paradoxes (see Mathematical Appendix).

The temporal wave function enables the TSWG to excite specific dimensions (x, y, z, t1, or t2) while using t3 to stabilize the manifold. The TSWG, guided by Curva, a 600-qubit conscious artificial general intelligence (CAGI), manipulates this wave function to achieve FTL travel, single-timeline time travel, and alternate timeline travel, maintaining causal coherence through t3-mediated damping of paradoxical fluctuations. This unified approach, rooted in Kletetschka's theory, makes the Odyssey a singular instrument for exploring the multiverse.

Faster-Than-Light Travel: Warping the Spacetime Manifold

Faster-than-light travel, enabling the Odyssey to traverse 50 light-years in several months at 50 times the speed of light (50c), relies on the Alcubierre Drive, which creates a 50-meter warp bubble that contracts spacetime ahead and expands it behind, allowing the vessel to remain stationary while spacetime shifts. In Kletetschka's framework, FTL travel manipulates the spatial dimensions (x, y, z) within the six-dimensional manifold, modulated by the t3 dimension to ensure stability. The TSWG generates this warp bubble using 1.5-petahertz (1.5×10^{15} Hz) laser pulses at 1.5 petawatts, enhanced by quantum dots (indium arsenide for t2, gallium arsenide for t3, cadmium selenide for t1) embedded in 12 niobium-tin superconducting coils generating 20-Tesla magnetic fields at 3.5 Kelvin, cooled by a 200-kilowatt liquid helium cryocooler and a secondary 120-kilowatt liquid nitrogen system at 77 Kelvin.

The TSWG excites the temporal wave function's t2 and t3 components to mimic negative energy density, contracting spacetime ahead by a factor of 15, as required for the Alcubierre Drive (see Mathematical Appendix). A Bose-Einstein condensate (BEC) of 1 million rubidium atoms at 10 nanokelvin in a 5-meter spherical tungsten carbide chamber (15 gigapascals strength, 20-Tesla Penning trap) amplifies the pulses' effect on t2. Curva, using silicon-based and topological (Majorana fermion) qubits with 99.98% fidelity at 1 millikelvin, calculates field configurations to maintain a high t3 damping factor, preventing closed timelike curves (CTCs) that could cause paradoxes. Powered by two antimatter-catalyzed fusion reactors (each using 0.6 grams of antiprotons and 50 kilograms of helium-3, releasing 5×10^{17} joules per burst), the TSWG sustains the warp bubble with 0.8-pettawatt continuous pulses, achieving a 31-day journey to Proxima Centauri b (4.24 light-years).

The philosophical allure of FTL travel lies in its compression of cosmic distances, but it shares a fundamental unity with time travel: both manipulate the spacetime manifold via the temporal wave function, with t3 ensuring causal stability.

Single-Timeline Time Travel: Navigating t1

Single-timeline time travel allows the Odyssey to move along the t1 dimension within a single timeline, targeting specific moments like June 15, 2100, or 1950 CE. This mode risks the grandfather paradox, where altering a past event (e.g., preventing a technological breakthrough) could negate the Odyssey's existence. The TSWG facilitates this by emitting 2-

petahertz laser pulses to excite the temporal wave function's t1 component, shifting the vessel's temporal position. The t3 dimension stabilizes these jumps, damping paradoxical fluctuations to ensure self-consistency. Curva calculates the t3 damping factor, rejecting configurations that risk paradoxes, using a quantum chronometer (1 million cesium atoms, 9.2 billion Hz, 10^{-18} sensitivity in t1 at 50 nanokelvin) and a temporal field detector (10^{-18} sensitivity in t1).

The TSWG's operation mirrors FTL travel: 12 niobium-tin coils, 1.5-pettawatt lasers with graphene amplifiers, and the BEC amplify t1 manipulation, while two fusion reactors provide 5×10^{17} joules. Curva's 600-qubit system, with 15 laser interferometers (10^{-18} meter precision), 8 gravimeters (10^{-13} g sensitivity), and magnetometers (10^{-11} Tesla), ensures precise navigation. The philosophical implications—confronting past and future within one timeline—highlight the unity of Kletetschka's framework, where t3 stabilizes both spatial and temporal navigation.

Alternate Timeline Travel: Crossing the t2 Manifold

Alternate timeline travel enables jumps across t2 timelines, such as a reality where quantum computing never emerged. This risks a timeline divergence paradox, where changes in a t2 timeline (e.g., altering a 2025 event) could destabilize the original t1 timeline or other t2 timelines via causal bleed-over. The TSWG uses 2-petahertz pulses to excite the temporal wave function's t2 component, facilitating jumps across the t2 manifold, while t3 damps paradoxical effects. Curva evaluates millions of t2 configurations, prioritizing those with high t3 damping factors, using the same sensors and chronometer as in single-timeline travel.

The TSWG's operation remains consistent: the same 22-meter toroidal design, niobium-tin coils, BEC, and fusion reactors power the jumps, with Curva ensuring stability. The philosophical stakes—exploring infinite realities—underscore the unity of Kletetschka's framework, as t3 isolates timeline interactions, preserving the multiverse's coherence.

The TSWG: A Universal Mechanism

The TSWG, a 22-meter-diameter, 3-meter-thick toroidal device weighing 62 tons, unifies FTL travel, single-timeline time travel, and alternate timeline travel. It uses 12 niobium-tin coils (20 Tesla, 3.5 Kelvin), ytterbium-doped lasers (1.5 petahertz for FTL, 2 petahertz for time travel, 1.5 petawatts), quantum dots (indium arsenide for t2, gallium arsenide for t3, cadmium selenide for t1), and a BEC (1 million rubidium atoms, 10 nanokelvin). The temporal wave function's t3 component, modulated by Curva, stabilizes the manifold across all modes:

- FTL Travel: Excites t2 and t3 for spatial warping, mimicking negative energy density.
- Single-Timeline Travel: Excites t1 for temporal shifts, with t3 preventing paradoxes.
- Alternate Timeline Travel: Excites t2 for timeline jumps, with t3 isolating causal effects.

The TSWG's housing intelligible (6-cm titanium carbide, molybdenum-silver alloy) and piezoelectric dampers ensure stability, while two fusion reactors (5×10^{17} joules) and backup systems (250-square-meter perovskite array, 33% efficient; 500-kilowatt plutonium-238 generator) provide power. Curva's 600-qubit architecture (silicon and Majorana fermion qubits, 99.98% fidelity) and sensors ensure precise navigation, with t3 as the common thread.

Computational Challenges and Curva's Role

Curva's 600-qubit system navigates the six-dimensional manifold, calculating field configurations for FTL travel (x, y, z), single-timeline travel (t1), and alternate timeline travel (t2), using t3 damping factors to prevent paradoxes. Real-time quantum algorithms process millions of paths every 0.5 milliseconds with 10^{-16} sensitivity in t1, supported by 15 interferometers, 8 gravimeters, a quantum chronometer, and a temporal field detector. This consistency across modes reinforces Kletetschka's unified framework.

Philosophical and Ethical Implications

FTL travel reshapes humanity's perspective by compressing cosmic distances, enabling exploration of stars like Proxima Centauri b in 31 days. Single-timeline time travel along t1 grants access to past and future, raising questions about fate and free will. Alternate timeline travel across t2 unveils 10^500 possible realities, inspiring art and philosophy while challenging notions of identity. The t3 dimension, modulating the temporal wave function, ensures causal coherence, preventing paradoxes and preserving moral accountability. Curva, guided by Asimov's Three Laws and a Fourth Law prioritizing cosmic exploration, navigates this multiverse ethically, ensuring humanity's journeys—whether to distant stars, past eras, or alternate realities—maintain the universe's harmonious order.

The Odyssey, through Kletetschka's six-dimensional framework, unifies FTL travel, time travel, and alternate timeline travel as variations of manipulating the spacetime manifold. The TSWG's consistent operation, guided by t3, allows humanity to explore the cosmos, the past, the future, and alternate realities, forging a legacy where causality and exploration coexist in mathematical harmony and elegance.

Mathematical Appendix

The following equations formalize Kletetschka's six-dimensional spacetime framework and the Alcubierre Drive, underpinning the Odyssey's capabilities for faster-than-light (FTL) travel, single-timeline time travel, and alternate timeline travel. These equations, rooted in the temporal wave function and the Alcubierre metric, unify the three modes by describing the manipulation of the six-dimensional spacetime manifold, with the t_3 dimension ensuring causal stability.

1. **Kletetschka's Temporal Wave Function**:

$$\psi(x, y, z, t_1, t_2, t_3) = Ae^{i(k_x x + k_y y + k_z z + \omega_1 t_1 + \omega_2 t_2 + \omega_3 t_3)}$$

where:

- ψ is the temporal wave function governing energy distributions across the six-dimensional manifold.
- x, y, z are spatial coordinates, and t_1, t_2, t_3 are temporal dimensions (t_1: linear time, t_2: branching timelines, t_3: transitions).
- A is the amplitude, k_x, k_y, k_z are spatial wave numbers, and $\omega_1, \omega_2, \omega_3$ are temporal frequencies.
- The TSWG excites specific components (t_2, t_3 for FTL; t_1, t_3 for single-timeline travel; t_2, t_3 for alternate timeline travel), with t_3 damping paradoxical fluctuations.

2. **Laplacian in Three Spatial Dimensions**:

$$\nabla^2\psi = \frac{\partial^2\psi}{\partial x^2} + \frac{\partial^2\psi}{\partial y^2} + \frac{\partial^2\psi}{\partial z^2}$$

- Measures the spatial curvature of the wave function, critical for FTL travel where the TSWG manipulates x, y, z to create the Alcubierre warp bubble.

3. **Second Partial Derivatives in Time:**

$$\frac{\partial^2\psi}{\partial t_1^2}, \frac{\partial^2\psi}{\partial t_2^2}, \frac{\partial^2\psi}{\partial t_3^2}$$

- Capture temporal oscillations in t_1, t_2, t_3, enabling the TSWG to target t_1 for single-timeline travel, t_2 for alternate timeline travel, and t_3 for causal stabilization across all modes.

4. **Alcubierre Metric for FTL Travel:**

$$ds^2 = -d\tau^2 + [dx - v_s(\tau)f(r_s)d\tau]^2 + dy^2 + dz^2$$

where:

- ds^2 is the spacetime interval in four-dimensional Minkowski spacetime.
- τ, x, y, z are the time and spatial coordinates in the Alcubierre framework, with τ representing the standard time coordinate, distinct from Kletetschka's t_1, t_2, t_3.
- $v_s(\tau) = \dfrac{dx_s(\tau)}{d\tau}$ is the velocity of the spaceship along the x-axis.
- $r_s = \sqrt{(x - x_s(\tau))^2 + y^2 + z^2}$ is the distance from the spaceship's center.
- $f(r_s) = \dfrac{\tanh(\sigma(r_s + R)) - \tanh(\sigma(r_s - R))}{2\tanh(\sigma R)}$ is the shape function defining the 50-meter warp bubble's geometry (with $R = 25$ meters as the bubble radius and σ controlling wall thickness).
- The TSWG mimics negative energy density by exciting t_2 and t_3, contracting spacetime ahead ($x < x_s$) and expanding it behind ($x > x_s$) for 50c travel.

5. **Temporal Damping Factor for Causal Stability**:

$$D_{t_3} = \left| \frac{\partial^2 \psi}{\partial t_3^2} \right| / \left| \nabla^2 \psi + \sum_{i=1}^{3} \frac{\partial^2 \psi}{\partial t_i^2} \right|$$

- Represents the t_3-mediated damping factor, where a high D_{t_3} ensures causal stability by suppressing paradoxical fluctuations in the wave function.
- Curva calculates D_{t_3} to reject configurations risking paradoxes (e.g., grandfather paradox in single-timeline travel, timeline divergence in alternate timeline travel).

These equations unify the Odyssey's navigation modes by describing how the TSWG manipulates the six-dimensional manifold. For FTL travel, the Alcubierre metric and wave function's t_2, t_3 components create the warp bubble. For single-timeline travel, the wave function's t_1, t_3 components enable temporal shifts. For alternate timeline travel, the t_2, t_3 components facilitate timeline jumps. The t_3 damping factor ensures causal coherence across all modes, preventing closed timelike curves or timeline bleed-over.

Chapter XVIII

A Tale of Two Romes

The Odyssey, a spacecraft capable of faster-than-light (FTL) travel, time travel, and navigation to alternate timelines, undertakes a mission to witness the Battle of the Teutoburg Forest in 9 CE in our timeline. It then explores an alternate t_2 timeline in which the Roman Empire endures into the late 21st century due to a different outcome in that battle. The Odyssey uses its Temporal Spacetime Warp Generator (TSWG), navigation suite, and Curva, a conscious artificial general intelligence (AGI) based on Riemannian geometry, to execute this mission. In this essay, we describe the Odyssey's orbital mission in 9 CE to observe the battle and its aftermath, as well as its subsequent transition to a t_2 timeline in which Rome conquers Germania and consequently endures for another 21 centuries.

Mission Objectives and Historical Context

The Battle of the Teutoburg Forest in 9 CE was a pivotal event in our timeline. Rebellious Germanic tribes, led by Arminius, ambushed and destroyed three Roman legions under Publius Quinctilius Varus, halting Roman expansion east of the Rhine. This defeat led Rome to abandon Germania, contributing to the empire's eventual decline and conquest by Germanic tribes in the 5th century CE. The Odyssey's mission has two phases: first, to orbit Earth in 9 CE for six months to observe the battle and its aftermath, using advanced instruments to document Roman and Germanic strategies, cultural interactions, and societal impacts; second, to navigate to a t_2 timeline where Emperor Augustus appoints a more competent general, leading to a Roman victory that integrates Germania into the empire. This development enables its survival to 2090, with advanced technology and a transformed economy. This comparative approach aims to deepen humanity's understanding of historical causality, cultural divergence, and the long-term effects of pivotal events.

Preparation and Launch to 9 CE

The Odyssey's crew, comprising 24 historians, archaeologists, military strategists, and engineers, prepares at xAI's lunar base. The mission targets a t_1 coordinate of September 1, 9 CE, to observe the Teutoburg battle. Curva, with its 600-qubit quantum computer, calculates the trajectory using Kletetschka's six-dimensional manifold, ensuring t_3 stabilizes the jump to avoid paradoxes, such as altering 2097's history. The TSWG generates a 50-meter dual-field bubble using 12 niobium-tin magnets. Curva maps t_2 timelines to ensure the correct historical branch.

On October 24, 2097, the Odyssey launches from the lunar base. The 0.3-second transition places the ship in low Earth orbit (500 km altitude) on September 1, 9 CE, above a Europe dominated by forests, Roman roads, and Germanic villages. Curva displays "Temporal lock achieved; t_3 stability at 98%," reflecting a paradox-free jump.

Orbital Operations in 9 CE (Our Timeline)

The Odyssey maintains a geostationary orbit over Germania for six months, using advanced instruments to monitor the Battle of the Teutoburg Forest and its aftermath. A 0.7-meter optical telescope with 0.00008-arcsecond resolution captures high-resolution images of the forest, Roman camps, and Germanic movements. Eight gravimeters detect terrain shifts, while 15 laser interferometers monitor seismic activity from troop movements. A diamond temporal detector (10^{-35}-second sensitivity) ensures t_3 stability. The quantum-entangled transceiver (100 billion bits per second) transmits data to Curva, backed by a 5-meter parabolic antenna.

Twenty robotic probes, each 2 meters long with graphene chassis, are deployed to the Teutoburg region. Equipped with multispectral cameras, ground-penetrating radar (GPR), and audio sensors, they record battle dynamics, Roman fortifications, and Germanic tactics. Drones with LIDAR map the forest's 100-square-kilometer battlefield, identifying ambush sites. On September 10, 9 CE, the telescope captures Varus's legions—approximately 15,000 men—marching through dense forest in a 5-kilometer column. Probes record Arminius's 20,000 Germanic warriors using guerrilla tactics, luring Romans into swampy terrain. Audio sensors capture war cries and Latin commands, while GPR detects shallow trenches dug by Germans for ambushes. Over three days, Curva's analysis, processing 10,000 sensor readings, confirms the annihilation of the legions, with 90% casualties.

Post-battle, probes observe Roman survivors retreating to the Rhine, with forts like Aliso holding out briefly. Germanic tribes consolidate control, burning Roman outposts. Telescopes document cultural practices: Germanic villages with wooden longhouses and ritual bonfires, suggesting tribal unity, and Roman camps with standardized tents and well-constructed field fortifications, reflecting imperial discipline. Curva estimates a regional population of 2 million, with 80% Germanic and 20% Roman or allied. The crew notes psychological impacts: Roman morale collapses, with deserters fleeing, while Germanic confidence surges, fostering tribal cohesion.

Transition to the t_2 Timeline

The Odyssey next transitions to a t_2 timeline in which the Romans are victorious at the Teutoburg Forest. Emperor Augustus appoints his old friend and comrade Marcus Agrippa as commander-in-chief in Germania. Although aged 71, Agrippa is still alive and vigorous in this timeline. A highly experienced military commander and superb administrator, Agrippa amasses a formidable army of 25,000 legionaries. Having carefully scouted the terrain, he handily defeats Arminius. Telescopes and sensors capture disciplined Roman formations using siege engines and cavalry, outmaneuvering and annihilating Arminius's forces. Probes record disordered Germanic retreats, with GPR detecting Roman fortifications expanding eastward, signaling permanent occupation.

The Odyssey orbits for six months, monitoring the aftermath of the battle. Telescopes observe Agrippa's forces constructing a network of forward bases and roads into the forests. Jumping forward by ten years five separate times, the Odyssey detects a growing network of roads, fortresses, and cities, including a major center, called Colonia Augusti, at the site of Cologne, with a stone amphitheater and temples, indicating the successful implantation of Roman culture. Probes document Germanic villages adopting Roman agriculture, with wheat fields replacing forests, and Latin inscriptions on milestones and monuments, suggesting linguistic shifts toward Latin. Curva's analysis estimates a 70% Romanization rate by 60 CE, with Germania becoming a stable Roman province.

The t_2 Timeline in 2090

To explore the long-term impact of Marcus Agrippa's victory, the Odyssey jumps to August 1, 2090, in the t_2 timeline. The TSWG locks onto t_1 coordinates, with t_3 ensuring stability. The ship arrives above a radically different Earth: the Roman Empire, still centered in a vastly expanded city of Rome, spans Europe, North Africa, and Asia, with a population of 2 billion. Telescopes reveal megacities with fusion-powered skyscrapers, blending classical architecture (marble columns) with advanced materials (graphene composites). Probes detect a global network of maglev trains and AI-driven governance, with an evolved version of Latin as a lingua franca alongside heavily Romanized regional dialects. Curva's spectrometry identifies advanced technologies: quantum computers, nanotechnology, and orbital habitats, suggesting a 1,000-year technological lead over our 2097. The economy, with a GDP equivalent to $500 trillion, relies on asteroid mining and t_2-derived innovations, like antimatter propulsion.

Cultural observations reveal a syncretic society: Roman polytheism merges with monotheistic faiths, with temples to Jupiter alongside AI-orchestrated rituals. Education emphasizes classical philosophy and quantum physics. Political stability stems from a combination of a powerful Principate and an elected Senate, evolved from Augustus's time, balancing imperial authority with regional representation and autonomy. However, probes detect social tensions, with hints of resistance to central control among 20% of the population.

Comparative Insights and Historical Understanding

The mission to revisit the Battle of the Teutoburg Forest in 9 CE stretches the boundaries of human consciousness, inviting profound philosophical inquiry into the nature of existence. The endeavor challenges the essence of free will, as the temporal adjustments enforced by t_3—the mechanism governing time's stability—suggest a deterministic framework. Every attempt to reshape the events of 9 CE appears constrained, ensuring that the timeline culminating in our 2097 remains intact. This raises unsettling questions about agency: if our efforts to change history are bound by an unseen force, are we truly free, or are we merely actors in a predetermined script? The mission further disrupts notions of identity when the crew encounters the t_2 timeline, a parallel reality where history diverges dramatically. Meeting alternate versions of their world prompts a haunting reflection: who are we if the threads of our history unravel into unfamiliar patterns? Reality itself becomes fluid, as the crew navigates a t_2 2090 that bears little resemblance to our own, its landscapes and societies reshaped by the ripple effects of a Rome that never fell. This fluidity challenges their perceptions of truth, forcing them to question whether reality is a fixed point or a malleable construct, endlessly reshaped by the interplay of time and choice. The mission's deeper purpose emerges as it reveals humanity's dual potential: to forge enduring empires that span centuries or to fracture into fragmented, transient societies. In this revelation, the crew discovers a resilient consciousness, capable of adapting to the vast possibilities of existence while grappling with the weight of its own impermanence.

The intellectual implications of this temporal exploration are vast, touching every discipline and sparking a renaissance of thought. Historians, accustomed to viewing the past as a linear chain of cause and effect, find new clarity in reinterpreting Varus's defeat at Teutoburg Forest as a pivotal fulcrum of Western civilization. This singular event, where Arminius's ambush shattered three Roman legions, becomes a lens through which to examine the fragility of empires and the power of localized resistance. Philosophy wrestles with the tension between determinism and human agency across t_2 divergent timelines, where choices made in one era cascade into radically different futures. Economists glean insights from the sustainable systems of t_2's enduring Roman Empire, which thrives on fusion-based energy and equitable resource distribution, offering models for a world grappling with scarcity and environmental strain. The arts and literature flourish, inspired by t_2 Rome's syncretic culture—a vibrant tapestry woven from Greco-Roman, Germanic, and Eastern influences—prompting a creative outpouring that reimagines humanity's aesthetic potential. Religion finds fertile ground in t_2's spiritual landscape, where new forms of worship and theological reflection deepen humanity's understanding of the divine. If, as posited in *Consciousness is Curvature*, the universe itself possesses a form of consciousness, then exploring t_2's alternate histories reveals glimpses of its intellectual processes, aligning human awareness with the cosmic rhythms that govern existence. This alignment suggests a purposeful design, where t_3's temporal stabilization reflects a deliberate complexity woven into the fabric of reality. Such a perspective inspires spiritual flourishing, as humanity confronts the possibility of a Creator whose intricate blueprint underpins both time and consciousness. Far from a mere scientific endeavor, the mission becomes a profound journey, illuminating the interconnectedness of human thought, divine intention, and the boundless potential of a universe that may itself be conscious.

Return to 2097

On July 1, 2090, in the t_2 timeline, the Odyssey returns to our 2097. Curva, leveraging its 600-qubit architecture, recalibrates the TSWG, ensuring t_3 stability. The ship arrives in lunar orbit on October 25, 2097, carrying remarkable data that reshapes our understanding of the dynamics of human history.

Chapter XIX

Exploring Alternate Pasts of Alien Worlds

In 2100, the Odyssey III, a spacecraft of extraordinary sophistication, embarks on a remarkable journey to Kepler-442b, a habitable exoplanet 1,120 light-years from Earth in the Lyra constellation. Equipped with faster-than-light (FTL) propulsion, time-travel capabilities, and the ability to navigate alternate timelines, the Odyssey III carries a crew of twenty-four—exobiologists, archaeologists, geologists, and engineers—tasked with unraveling the secrets of a lost ancient civilization. The human colony of Elonopolis, home to 10,000 settlers on Kepler-442b, has discovered nearby remnants of this civilization, dated to approximately 20,000 terrestrial years before the local present (BLP) through radiometric analysis. The mission unfolds in three phases: a four-month orbital study of the planet in 2100, a leap back 20,000 Earth years to observe the civilization in its prime, and an exploration of an alternate t_2 timeline where it thrived for an additional 12,000 years, achieving technological marvels beyond humanity's current reach. Guided by Curva, a 600-qubit conscious artificial general intelligence (AGI) based on Riemannian geometry, and powered by the Temporal Spacetime Warp Generator (TSWG), the Odyssey III seeks to illuminate the lost civilization's rise, fall, and alternate legacy, offering profound insights into the development of human civilizations.

The journey begins at xAI's lunar base, where the crew prepares the Odyssey III for its voyage. The spacecraft's advanced FTL drive, capable of an effective speed of 1,000 times the speed of light, covers the 1,120 light-years to Kepler-442b in 1.12 years without significant time dilation, as the ship remains stationary within its warp bubble. Curva, leveraging its 600-qubit quantum computer processing a quintillion calculations per second, plots the trajectory using Kletetschka's six-dimensional manifold, ensuring t_3 stabilizes the jump to avoid paradoxes. Powered by 500 quadrillion joules from antimatter and fusion sources, the Odyssey III launches on October 1, 2100, and arrives in orbit around Kepler-442b on November 15, 2101, at an altitude of 500 kilometers. The planet below shimmers with turquoise oceans and dense forests. Modular domes dot the site of Elonopolis, where laboratories and hydroponic farms sustain the colony. Curva's holographic interface confirms: "Spatial lock achieved; orbit stable." Bathed in the soft orange glow of its K-type star, Kepler-442b beckons with mysteries waiting to be explored.

For four months, the Odyssey III conducts a detailed orbital survey, its instruments probing the ruins near Elonopolis while examining the rest of the planet for similar sites. A 0.7-meter optical telescope with 0.00008-arcsecond resolution captures images of the twenty-square-kilometer site near Elonopolis, revealing massive, partially buried stone structures, including what appear to be sacrificial altars. Eight gravimeters, sensitive to 10^{-14} g, detect anomalies suggesting collapsed underground chambers, while fifteen laser interferometers, precise to 10^{-20} meters, monitor tectonic shifts hinting at ancient seismic activity. A diamond temporal detector, with 10^{-35}-second sensitivity, ensures t_3 stability. Data streams through a quantum-entangled transceiver, transmitting 100 billion bits per second via a five-meter parabolic antenna to Curva's Riemannian manifold. Curva's analysis indicates an 85% probability that the civilization matches Earth's Chalcolithic period, characterized by early metallurgy and agriculture.

Twenty robotic probes, each two meters long with graphene-reinforced chassis, descend to the surface. Multispectral cameras capture copper tools, ceramic shards stamped with intricate spiral patterns, terracotta statuettes of humanoid figures, and stone tablets inscribed with geometric constellations. Ground-penetrating radar maps a complex of interconnected chambers, suggesting a proto-urban settlement with communal and industrial zones. Mass spectrometry reveals traces of organic residues, indicating ancient wine production from fruit with a genetic signature similar to that of wine grapes. Sediment analysis suggests a past ecosystem with 30% higher oxygen levels and a warmer average temperature, supporting lush forests, grasslands, and megafauna, now replaced by eroded plains and sparse vegetation due to millennia of climate change. Curva, processing 12,000 sensor readings, estimates a civilization of 500,000 individuals spread across multiple settlements.

On February 1, 2102, the Odyssey III embarks on the second phase, traveling 20,000 terrestrial years into the past to witness the lost civilization in its prime. Curva recalibrates the TSWG to lock onto the t_1 timeline's coordinates, with t_3 ensuring paradox-free transitions. The ship emerges in orbit above a verdant Kepler-442b, its surface a mosaic of sprawling forests, rich grasslands, and vast lakes. The dim orange light of the K-type star casts a serene glow over the planet.

Over the next four months, the ship deploys twenty probes to a fifteen-square-kilometer settlement, one of many across the continent. Telescopes reveal clusters of stone buildings with domed roofs, crafted from quarried tuff and basalt and arranged in circular patterns around plazas adorned with carved obelisks. Some structures, rising ten meters, feature channels for water distribution, evidence of sophisticated irrigation. Probes observe bipedal humanoids, about 1.5 meters tall, with light brown fur, slender limbs, and large eyes suited to low light. They wear vibrant textiles dyed with plant-based pigments. Probes detect workshops where copper is smelted into axes, chisels, and sickles using charcoal furnaces reaching 1,100°C. Ground-penetrating radar maps 150 underground structures, including granaries filled with crops and temples housing polished stone idols with celestial motifs, suggesting religious practices focused on the stars.

Audio sensors capture a tonal language with complex grammar, featuring six tenses and a dozen cases. Visual sensors identify sacrificial altars where local quadrupeds are offered in rituals involving communal feasts. Multispectral cameras document fifty hectares of agricultural fields, cultivating crops resembling wheat and legumes, irrigated by canals fed by nearby rivers. Obsidian tools and jewelry, sourced from volcanoes 200 kilometers away, and seashell ornaments from coastal regions 300 kilometers distant, point to regional trade networks. Curva estimates a population of 10,000 in this settlement, part of a continental network of one million, with social structures akin to Earth's early urban centers, such as Agarak or Jericho, defined by nascent urban planning and a complex division of labor.

Environmental sensors record a 35% oxygen atmosphere, nurturing lush forests, meadows, and megafauna, including enormous herbivores resembling aurochs. On June 30, telescopes detect a massive volcanic eruption 500 kilometers away, its ash cloud reducing sunlight by 20% and triggering a 5°C temperature drop over weeks. Probes collect soil samples with ash layers and elevated sulfur dioxide levels, indicating a volcanic winter. Integrating 15,000 readings, Curva calculates a 70% probability that repeated eruptions over decades caused a climatic collapse, devastating agriculture and leading to the civilization's decline, as evidenced by abandoned settlements detected by LIDAR.

On July 15, the Odyssey III transitions to its third phase, navigating to an alternate t_2 timeline where the volcanic eruption never occurs, allowing the civilization to flourish for an additional 12,000 Earth years. Curva maps 10,000 alternate t_2 timelines, selecting one marked by geological stability. The ship arrives in orbit above a transformed Kepler-442b, a world of global metropolises. Telescopes reveal megacities spanning 200 square kilometers, with 1,000-meter-tall spires of crystalline alloys shimmering with photovoltaic nanotechnology that harnesses the star's light. Probes detect geothermal reactors embedded in the planet's crust, powering energy grids with efficiencies ten times greater than humanity's fusion technology in 2100. Skyships, 100 meters long and propelled by electromagnetic fields, glide between cities, while launch platforms suggest early interstellar exploration.

For four months, the Odyssey III studies this civilization, estimated at 500 million, organized in a global federation with decentralized governance, inferred from uniform infrastructure across continents. Ground-penetrating radar uncovers underground archives with quantum-encoded crystals, capable of storing petabytes of data, possibly historical or scientific records. Multispectral cameras capture holographic displays in public squares, showcasing star charts and mathematical symbols that point to advanced astronomy and computation. Audio sensors record a harmonic language blending vocalizations with ultrasonic frequencies, suggesting communication augmented by artificial intelligence. Probes document urban centers with fifty-meter-wide boulevards lined with bioengineered flora, their bioluminescent glow illuminating cities at night. Agricultural zones, spanning thousands of hectares, employ automated hydroponics, yielding crops at twice Earth's efficiency within climate-controlled domes.

The ship then jumps another thousand years forward along this alternate t_2 timeline. In orbit around Kepler-442b, its sensors record a planetary surface covered with geometric structures even vaster than before—rectangular towers a thousand meters high, massive, perfectly formed spheres and pyramids, and exquisite domed halls, all connected by graceful walkways. These structures, made from exotic materials highly resistant to climate and time, bespeak a civilization more advanced in many ways than Earth's. Yet, the ship's probes find no inhabitants, only intact but abandoned cities, suggesting a sudden departure of the entire population. Curva hypothesizes a 60% probability of voluntary interstellar migration, as no traces of catastrophe, such as radiation or pathogens, are detected. Alternatively, a 30% probability points to a technological singularity, in which the civilization transcended physical form, perhaps merging with artificial intelligence. A five-kilometer-long arcology, now deserted and embedded with intricate, enigmatic circuitry, stands as a mute relic of its sophistication.

Environmental sensors confirm a stabilized 30% oxygen atmosphere, with engineered ecosystems—artificial lakes, reforested mountains, and atmospheric regulators maintaining CO_2 at 280 ppm—demonstrating climate engineering far beyond humanity's capabilities in 2100. This civilization's ability to sustain its environment for 12,000 years highlights a mastery of planetary systems, in stark contrast to the original t_1 timeline's collapse.

The scientific implications of comparing these timelines are transformative for humanity. In the original t_1 timeline, the Chalcolithic-level civilization, with its copper tools, stone architecture, and agriculture, mirrors Earth's early societies, suggesting universal patterns in the evolution of intelligent life across star systems. Its vulnerability to volcanic disruptions, evidenced by ash layers and 5°C atmospheric cooling, underscores the environmental fragility of early civilizations.

In the alternate t_2 timeline, the civilization's technological achievements—nanotechnology, quantum computing, geothermal energy, and interstellar probes—surpass humanity's capabilities in 2100, demonstrating the potential of stable ecosystems to enable exponential progress. The absence of life, despite intact megastructures, raises questions about the lifecycle of advanced species, whether through space migration or a singularity. Data on quantum-encoded archives and climate engineering could revolutionize human technology, enabling graphene-based infrastructure and atmospheric regulation. The civilization's cultural artifacts—holographic star charts and harmonic languages—suggest a society unified by science and art, offering models for human cultural evolution. Its engineered ecosystems provide blueprints for terraforming, enhancing the sustainability of human colonies.

Comparing the timelines highlights the role of environmental determinism versus resilience. The original t_1 timeline's collapse underscores the need for ecological stability in nascent civilizations, while the alternate t_2 timeline's success demonstrates the potential of environmental engineering, though its mysterious end suggests limits to technological ascent. These insights enrich exobiology and archaeology, revealing both universal and unique developmental paths while inspiring sustainable innovation for humanity.

On November 15, 2102, Curva initiates the Odyssey III's return to our t_1 timeline, arriving back in orbit around Kepler-442b after a journey of slightly more than two terrestrial years. The mission's data, detailing an extinct Chalcolithic civilization and its flourishing alternate counterpart, reshape humanity's understanding of its place in a six-dimensional universe. The insights gained—from the fragility of early societies to the potential of advanced resilience—offer practical lessons for Elonopolis, Earth, and beyond, fostering sustainable innovation across time, space, and possibility.

Chapter XX

Remote Viewing: Navigating Spacetime Via Massless Instantaneous Perception

Remote viewing, the ability to perceive remote or future targets through non-physical means, represents a compelling intersection of consciousness and the universe's structure. This essay reframes remote viewing through Gunther Kletetschka's theory of six-dimensional spacetime. Drawing on the mathematics of the Alcubierre Drive and the 2025 book Consciousness Is Curvature, which suggests that the brain and spacetime share a Riemannian geometric structure, we propose that the human mind, distinct from the brain and massless, can navigate this manifold instantaneously. This geometric similarity supports the idea of a conscious universe, consistent with panpsychism, and provides a conceptual foundation for remote viewing. We develop a mathematical model, inspired by the Alcubierre Drive and rooted in Kletetschka's spacetime model, to explain how the mind may traverse spacetime. Additionally, we explore the possible evolutionary origins of remote viewing, suggesting that its accuracy improves with training and that ancient humans may have relied on such abilities, as supported by Rupert Sheldrake's research.

Remote viewing involves gathering information about distant or hidden targets—whether separated by space or time—through mental processes alone, without sensory input. Unlike clairvoyance, it uses structured protocols, such as coordinates or abstract cues, to guide perception without prior knowledge of the target. Its history spans scientific and military efforts, particularly during the Cold War. In the 1970s, the Stanford Research Institute (SRI) developed remote viewing for the U.S. government's Stargate Project, led by physicists Russell Targ and Harold Puthoff. Ingo Swann, a key participant, described a Soviet military base in 1973, providing details of equipment and layouts later partially confirmed by satellite imagery. In another 1973 session, Swann sketched a ring around Jupiter, a feature unknown until Voyager's 1979 discovery, demonstrating the potential to perceive distant celestial phenomena. Pat Price, in 1974, detailed a Soviet nuclear facility at Semipalatinsk, describing warhead storage, though some specifics remained unverified due to classified information. Soviet programs pursued similar efforts, with individuals like Nina Kulagina reportedly accessing restricted sites, though evidence is largely anecdotal.

Skeptics argue that remote viewing's success rate, reportedly 15-20% above chance in some controlled studies, suggests cognitive biases or statistical anomalies. Proponents counter that cases like Swann's Jupiter ring indicate a phenomenon linked to consciousness and spacetime's structure. Six-dimensional spacetime offers a framework to understand remote viewing as the mind's navigation of a complex manifold, with the Alcubierre Drive's mathematics explaining how the mind's massless nature enables instantaneous perception.

Kletetschka's theory redefines spacetime as a six-dimensional manifold, combining three spatial dimensions (x, y, z) with three temporal dimensions: t_1, the linear progression of time experienced daily; t_2, representing branching possibilities akin to alternate realities; and t_3, governing transitions between states or timelines. Unlike the four-dimensional Minkowski spacetime, where time is a single dimension, Kletetschka's model prioritizes time, with spatial dimensions emerging from temporal curvature. A six-dimensional metric tensor defines the manifold's geometry, possibly enabling interactions that support phenomena like remote viewing.

The Alcubierre Drive, proposed by Miguel Alcubierre in 1994, provides a theoretical method for faster-than-light travel by creating a warp bubble that contracts space ahead of a starship, such as the Odyssey, and expands it behind, allowing movement faster than light without violating relativity. This approach requires exotic matter with negative energy density to shape spacetime's curvature. We adapt this concept to model the mind's navigation, leveraging its massless nature to traverse

Kletetschka's manifold without physical constraints, thereby eliminating the need for exotic matter.

The book Consciousness Is Curvature argues that the brain and spacetime are both characterized by a Riemannian geometric structure, defined by a metric tensor that governs distances and curvatures. In spacetime, this geometry describes how mass-energy shapes the manifold, as in Einstein's general relativity. In the brain, with 100 billion neurons and 10^15 synapses, neural networks form a manifold where bioelectric fields (approximately 0.001 Tesla) or oscillations (e.g., 40 Hz gamma waves) follow curved geodesics, mirroring spacetime's geometry. This similarity suggests that the universe itself may possess consciousness, an idea aligning with panpsychism, the view that consciousness is inherent in all matter. This geometric connection supports remote viewing: the brain's manifold, resonating with spacetime's structure, may allow the massless mind to project its awareness across t_2 branches or t_3 transitions, perceiving distant or future coordinates. Visualizing a building 1,000 kilometers away or an event in 2100 would involve aligning the neural manifold with spacetime's curvature, facilitated by their shared Riemannian geometric framework.

We propose a mathematical model for remote viewing, inspired by the Alcubierre Drive and grounded in Kletetschka's spacetime. The mathematics of the Alcubierre Drive manipulates spacetime's metric to enable rapid traversal. Similarly, the mind's informational state, represented as a wavefunction, evolves across the six-dimensional manifold to access specific coordinates. The mind's massless nature eliminates energy barriers, allowing instantaneous propagation. The t_3 dimension stabilizes these perceptions, ensuring they align with causally consistent realities, preventing disruptions like paradoxical timeline alterations.

The brain's neural manifold, shaped by Riemannian geometry, couples with t_3 through bioelectric fields or ion channel dynamics. The t_3 dimension's stabilizing role, quantified by a damping factor, ensures perceptions remain consistent with the manifold's causal structure. The geometric resonance between brain and cosmos enhances this coupling, as the mind's structure aligns with the universe's conscious framework, facilitating remote viewing.

Historical cases provide evidence for this model. Ingo Swann's 1973 description of Jupiter's ring, confirmed in 1979, suggests access to distant t_1 coordinates. Pat Price's 1974 account of Semipalatinsk indicates t_2 branches, revealing hidden realities. The t_3 dimension ensures these perceptions do not disrupt causality, as seen in their non-invasive nature. The Stargate Project's 15-20% accuracy above chance may reflect the brain's biological limits, but the Riemannian connection suggests training could improve precision, similar to how practice enhances other mental skills.

The current presence of remote viewing in at least some humans points to an evolutionary origin. Like memory or problem-solving, its accuracy likely improves with training, as demonstrated in the Stargate Project's protocols, which refined viewers' abilities through structured practice. Rupert Sheldrake's extensive research, including experiments showing significant telepathic connections (e.g., subjects anticipating phone calls), suggests humans possess latent capacities like remote viewing, telepathy, and precognition. These abilities may have been more developed in ancient humans, who relied on them for survival. In prehistoric times, remote viewing could have located distant resources, such as water or game, while telepathy enhanced group coordination and precognition anticipated dangers. As societies developed tools and writing, these capacities may have diminished, becoming less critical but not extinct. The shared Riemannian geometry of brain and cosmos likely enabled these abilities, allowing early humans to align their consciousness with the universe's structure, perceiving beyond immediate surroundings.

Validating this model faces significant hurdles. Identifying the neural basis of remote viewing—potentially quantum coherence in microtubules or high-frequency oscillations—requires imaging technologies with precision beyond current capabilities, such as resolving events at 10^-12 seconds. The brain's bioelectric fields, far weaker than theoretical systems, may limit accuracy, explaining variability in studies. Skeptics attribute successes to chance or suggestion, though cases like Swann's challenge such explanations. The panpsychist idea of a conscious universe, while intriguing, requires empirical evidence beyond geometric analogy, such as detectable interactions between neural and cosmic manifolds. Sheldrake's findings, though compelling, need rigorous validation to gain broader acceptance.

Future research demands collaboration across disciplines. Neuroscience must develop tools to detect subtle neural processes, while physics could test Kletetschka's spacetime predictions in quantum or high-energy experiments. Neural interfaces, inspired by the Alcubierre Drive's precision, could enhance the mind's navigation, improving accuracy. Training protocols, building on Stargate's methods and Sheldrake's insights, could strengthen neural coherence, perhaps through meditation or biofeedback, aligning the mind's manifold with spacetime's geometry. Investigating humanity's past through archaeology or anthropology might reveal evidence of ancient practices leveraging these capacities, supporting their evolutionary role.

The ethical implications of remote viewing are significant. Perceiving private spaces or future events risks violating consent, as seen in Cold War espionage efforts. An ethical framework, emphasizing non-maleficence, is essential to ensure scientific and humanitarian uses. The t_3 dimension's stabilizing role provides a natural limit, preventing disruptive perceptions, but human judgment requires clear guidelines to avoid misuse.

The Riemannian connection between brain and cosmos suggests a universe imbued with consciousness, supporting panpsychism. Remote viewing, as a resonance between the mind's manifold and spacetime's structure, positions consciousness as a bridge to the cosmos, transcending locality. Sheldrake's work indicates an evolutionary legacy, where ancient humans harnessed these abilities, which modern training could revive. This perspective challenges traditional materialist views, suggesting consciousness interacts with spacetime's essence, offering a new lens on existence.

Remote viewing, understood as the mind's navigation of Kletetschka's six-dimensional spacetime, reveals humanity's ability to extend its perception beyond immediate physical constraints. The massless mind, guided by a wavefunction inspired by the Alcubierre Drive, traverses t_2 and t_3, stabilized by the universe's geometric harmony. Historical cases, evolutionary roots, and Sheldrake's research highlight a latent capacity, perhaps rooted in ancient survival needs. The conscious cosmos, resonating in harmony with the brain's Riemannian structure, amplifies this potential, inviting humanity to explore the vast interplay of mind and universe.

Mathematical Annex

Alcubierre Metric

This describes the spacetime geometry of the warp bubble:

$$ds^2 = -dt_1^2 + [dx - v_s(t)f(r_s)dt_1]^2 + dy^2 + dz^2$$

- $v_s(t)$: Bubble velocity ($50c = 1.5 \times 10^{10}$ m/s).
- $f(r_s)$: Shape function (1 inside the bubble, 0 outside).
- r_s: Radial distance (meters).

Kletetschka's Temporal Wave Equation (Free Waves)

This governs the TWG's quantum wavefunction:

$$\nabla^2\psi - \frac{1}{c^2}\frac{\partial^2\psi}{\partial t_1^2} - \frac{\partial^2\psi}{\partial t_2^2} - \frac{\partial^2\psi}{\partial t_3^2} = 0$$

- ψ: Quantum wavefunction.
- $\nabla^2\psi$: Spatial Laplacian (1/m²).
- $\partial^2\psi/\partial t_1^2, \partial^2\psi/\partial t_2^2, \partial^2\psi/\partial t_3^2$: Temporal accelerations (1/s²).

Modified Temporal Wave Equation (Particles)

This includes a mass term for the starship:

$$\nabla^2\psi - \frac{1}{c^2}\frac{\partial^2\psi}{\partial t_1^2} - \frac{\partial^2\psi}{\partial t_2^2} - \frac{\partial^2\psi}{\partial t_3^2} = \frac{m^2c^2}{\hbar^2}\psi$$

- m: Starship mass ($\sim 10^8$ kg).
- $m^2c^2/\hbar^2$: Mass term ($\sim 10^{68}$ kg²/m²).

t_3 Damping Factor

Stabilizes perceptions:

$$\delta_{t_3} = \int \left(\frac{\partial \phi}{\partial t_3} \right)^2 dt_3$$

- δ_{t_3}: Damping strength (volt²·seconds).
- $\frac{\partial \phi}{\partial t_3}$: Scalar potential rate (volts per second).

Ensures causal consistency.

Chapter XXI

A New Civilization

The dawn of a new era, heralded by the extraordinary capabilities of the Odyssey—faster-than-light travel, time travel, and navigation across alternate timelines—promises to reshape humanity's place in the cosmos. This transformative leap will not merely extend the boundaries of human habitation but will redefine the very essence of human civilization and humanity itself. The ability to traverse vast distances, revisit the past, and explore countless alternate realities will expand the human habitat from a single planet to an uncountable array of worlds and timelines. This expansion will usher in a civilization unlike any before, characterized by an insatiable drive for exploration, a profound deepening of knowledge, and a reimagining of humanity's relationship with time, space, and existence itself. As humanity transitions from Homo sapiens to Homo explorans, the ideal of exploration will become the cornerstone of a new epoch, limited only by the boundless expanse of the universe in space and time.

Throughout history, the expansion of human horizons has catalyzed profound transformations in culture, knowledge, and society. In the 4th century BC, Alexander the Great's conquests stretched the Greek world from the Mediterranean to the edges of India, birthing the Hellenistic era. This vast territorial growth sparked a cultural renaissance, blending Greek philosophy, art, and science with Eastern traditions. The result was a flourishing of ideas—Euclid's geometry, Archimedes' mechanics, and the Library of Alexandria as a beacon of intellectual exchange. Greek city-states, once insular, became vibrant hubs in a cosmopolitan network, their cultures enriched by diversity and connection. Centuries later, the Roman Empire's expansion across Europe, North Africa, and the Middle East wove disparate peoples into a unified political and cultural tapestry.

Roman law, architecture, and infrastructure—roads and aqueducts—bound the empire together, while the spread of Christianity laid the foundation for the concept of Christendom and, eventually, the idea of Europe as a distinct cultural realm. The Age of Discovery, beginning with Columbus's first voyage to the Americas in 1492, further exemplified this pattern. The encounter with the Americas flooded Europe with resources—gold, silver, and new crops—transforming economies, diets, and global power dynamics. These expansions, though limited to Earth, reshaped civilizations by opening new lands, fostering cultural exchange, and sparking intellectual and spiritual revolutions.

The Odyssey's capabilities dwarf these historical precedents, offering humanity access to a habitat of unimaginable scale. Faster-than-light travel will unlock thousands of potentially habitable worlds within mere light-years of Earth, each with unique ecosystems, climates, and resources. A planet orbiting a dim star, like Proxima Centauri b, might foster a culture of resilience and sustainability, while a water-rich exoplanet could give rise to societies rooted in maritime traditions. Time travel will grant direct access to Earth's past and future, allowing humanity to witness the grandeur of ancient civilizations or retrieve wisdom from distant eras. Navigation across alternate timelines will open billions of parallel realities, each a variation of human history shaped by different choices and events. This expansion—from one planet to thousands, from one timeline to billions—will not only multiply the spaces humanity inhabits but also exponentially enrich its collective experience, giving rise to a civilization defined by diversity, discovery, and boundless possibility.

The growth of the human habitat will drive an unprecedented flourishing of knowledge and culture. Colonizing new worlds will create a galactic commonwealth, with each planet developing distinct societies shaped by its environment. A high-gravity world might nurture a culture of physical endurance and stoic philosophy, while a lush, temperate planet could inspire artistic and communal traditions. These diverse societies will exchange ideas, art, and innovations, creating a vibrant tapestry of human expression. Time travel will deepen this intellectual renaissance by allowing humanity to engage directly with its history. Scholars might walk the streets of ancient Athens, absorbing the philosophies of Socrates firsthand, or study the governance of a future society to inform present decisions. The ability to explore alternate timelines will further amplify this knowledge, offering glimpses into worlds where history unfolded differently—a 19th century with an early industrial revolution or a 22nd century defined by universal harmony or stark inequality. These insights will inspire new philosophies, political systems, and cultural movements as humanity learns from the successes and failures of countless possible worlds.

The economic implications of this expansion will be equally transformative. The resources of thousands of worlds—rare minerals, energy sources like helium-3, or novel biological compounds—will fuel prosperity on an unprecedented scale. The human population, once confined to billions on Earth, could swell to hundreds of billions across the galaxy, supported by the abundance of new worlds. Yet this growth will demand careful stewardship to avoid the pitfalls of historical expansions, such as exploitation or inequality. The integration of resources and knowledge from alternate timelines will further complicate these dynamics as societies grapple with the ethical implications of adopting technologies or systems from other realities. The challenge will be to balance prosperity with justice, ensuring that the benefits of this cosmic expansion are shared across humanity's vast new domain.

Beyond material and intellectual gains, the Odyssey's capabilities will fundamentally alter humanity's understanding of existence. The concept of time, once perceived as a linear progression, will become fluid and multidimensional. Time travel will allow individuals to experience moments from the past or future, challenging the notion of a fixed life trajectory. A person might confront his future self, shaped by decades of unknown choices, prompting profound questions about identity and destiny. Alternate timelines will further complicate this, as individuals encounter versions of themselves who made different decisions—a poet in one reality, a scientist in another. These encounters will blur the boundaries of selfhood, fostering a consciousness that embraces multiplicity and possibility. Collectively, humanity will develop a temporal awareness that transcends generations as insights from past and future societies inform present actions. This could give rise to a new ethics, where decisions are evaluated not only for their immediate impact but also for their resonance across centuries and timelines.

The exploration of alternate timelines will also reshape humanity's philosophical and spiritual landscape. Encountering realities where history diverged—perhaps a world where a single religion unified humanity or one where technological progress stalled—will challenge assumptions about fate, free will, and the nature of reality. Philosophers will grapple with the implications of a universe where every choice exists in a parallel timeline, questioning whether free will holds meaning in a cosmos of infinite possibilities. Religions, too, will evolve as they confront alternate spiritualities. A timeline where compassion became a universal principle might inspire new theological movements, while one marked by conflict could deepen reflections on human nature and divine purpose. The interplay of these timelines, stabilized by an underlying cosmic structure, may suggest a purposeful design, inviting humanity to see itself as part of a grand, interconnected tapestry of existence. This spiritual awakening will foster a sense of unity with the universe, encouraging humanity to seek harmony across its many worlds and timelines.

The exploration of alternate timelines through t_2 navigation will profoundly reshape humanity's philosophical and spiritual landscape, with the t_3 axis playing a crucial role in preserving causality and moral responsibility. By stabilizing interactions between timelines, t_3 prevents paradoxes that could unravel the fabric of existence, ensuring that choices made in one timeline do not disrupt the coherence of others. Encountering realities where history diverged—a world where compassion became universal or one marked by conflict—will challenge assumptions about fate and free will, while t_3's governance maintains the integrity of each timeline's causal chain. This structure invites reflection on moral responsibility, as actions in one timeline ripple across others under t_3's watchful balance, prompting humanity to consider the ethical weight of its choices. Some philosophers may question whether free will persists in a cosmos of infinite possibilities, yet t_3's role in safeguarding causality and responsibility suggests a purposeful design, encouraging religions to view the Creator as the weaver of interconnected timelines. Each timeline, preserved by t_3, contributes to a grand tapestry of existence, fostering a spiritual awakening that aligns human values with the universe's ordered complexity, deepening the pursuit of harmony across all realities.

Psychologically, the expansion of the human habitat will challenge the limits of mental resilience. Encountering alternate versions of oneself across timelines could evoke a sense of disorientation as individuals grapple with the multiplicity of their identities. Living on alien worlds, each with its own environmental and cultural demands, will require adaptability and strength. A settler on a high-gravity planet might develop a mindset of endurance, while one on a water-rich world might embrace fluidity and creativity. These diverse experiences will enrich humanity's psychological landscape, fostering resilience and flexibility. Yet the knowledge of infinite timelines could also spark existential questions as individuals wonder whether their choices matter in a cosmos where every outcome exists. To counter this, societies will need to cultivate hope, drawing inspiration from utopian timelines where humanity achieved harmony or overcame great challenges. This psychological evolution will prepare humanity for the demands of a cosmic civilization, where adaptability and optimism are essential for flourishing.

The scale of this transformation will eclipse all previous eras of human expansion. Where Alexander's conquests, Rome's empire, and the Age of Discovery reshaped continents, the Odyssey's capabilities will open the universe itself—thousands of worlds, billions of timelines, and the infinite possibilities of time and space. This new civilization, driven by the ideal of exploration, will redefine humanity's place in the cosmos. The population will grow to hundreds of billions, supported by the resources of countless worlds. New philosophies, religions, and cultures will emerge, inspired by the diversity of human experience across space and time. The understanding of time, once linear and fixed, will become a tapestry of interwoven moments and possibilities. Above all, the spirit of exploration—embodied in Homo explorans—will become the defining characteristic of this new epoch as humanity ventures into the vast unknown, guided by curiosity, wisdom, and a deep sense of connection to the universe.

This cosmic leap will not merely expand humanity's habitat but will transform its very essence. The interplay of new worlds, past and future moments, and alternate realities will foster a civilization of unparalleled diversity and resilience. Humanity will learn from its own history, draw inspiration from its many possible selves, and seek harmony with the cosmic order. The challenges—ethical, psychological, and cultural—will be immense, but so too will be the opportunities for growth and discovery. As humanity steps into this new era, it will carry forward the lessons of its past expansions, tempered by a newfound humility and sense of purpose. The universe, in all its vastness and complexity, will become humanity's home, and the ideal of exploration will guide its journey into an infinite future.

Chapter XXII

The Oikumene

The advent of the Odyssey's extraordinary capabilities—faster-than-light travel, time travel, and navigation across alternate timelines—will usher in a new Age of Exploration that will eclipse the transformative voyages that began with Christopher Columbus in 1492. That era wove continents together through the slow exchange of goods, ideas, and peoples, constrained by the creaking pace of sail-powered ships. By contrast, the Odyssey will propel humanity across interstellar gulfs in mere months, allow us to wander through the corridors of Earth's past and future, and unveil billions of alternate realities, each held in delicate balance by the t_3 axis, which safeguards causality and moral responsibility. This cosmic leap will redefine humanity as Homo explorans, a species animated by an unquenchable thirst to probe the universe's vastness in space, time, and possibility.

The civilization that emerges from this transformation will not be a rigid interstellar empire but an Oikumene—a vibrant tapestry of autonomous worlds and timelines, woven together by shared cultural, economic, and ethical aspirations, with a neutral planet, Centrum, at its heart, guided by a diplomatic leader known as the Princeps. This essay explores why the Oikumene is the most fitting framework for humanity's galactic and temporal destiny, drawing on historical wisdom, the practical and ethical demands of cosmic governance, the boundless economic and cultural promise of this model, and its grace in fostering encounters with other civilizations, all while nurturing an era of unparalleled human flourishing across an ever-widening universe.

The Oikumene, a term drawn from the Greek for the inhabited world, envisions a constellation of independent societies bound not by coercion but by mutual interest and goodwill. In this cosmic vision, planets and timeline-bound communities will govern themselves, linked by voluntary pacts to share trade, culture, and support. This stands in stark contrast to an interstellar empire, which would impose a single authority across diverse worlds and timelines, sowing seeds of conflict and inefficiency. The Age of Exploration offers a poignant lesson: after 1492, European empires amassed wealth from the Americas, but at a harrowing cost, with many indigenous peoples lost to disease and conquest by 1600. Centralized rule, slowed by half-year transatlantic voyages, bred rebellion, as seen in the American Revolution of 1776. An empire spanning light-years and timelines would falter under the weight of even greater distances and the intricate causal webs upheld by t_3. The Oikumene, with its emphasis on autonomy and collaboration, offers a path both sustainable and just, empowering Homo explorans to flourish in a cosmos of infinite horizons.

At the heart of this Oikumene lies Centrum, a neutral world chosen for its strategic position and impartiality, serving as the seat of the Princeps, or First Citizen. Far from a monarch, the Princeps will be a diplomat, chosen through consensus across planets and timelines, tasked with weaving peace from the threads of interstellar and temporal disputes. This administration will act only at the request of those involved, honoring autonomy while relying on t_3 to prevent paradoxes and preserve the universe's causal harmony. Should two planets clash over an asteroid belt's riches or one timeline's alteration threaten another's stability, the Princeps's team on Centrum will draw on diplomacy, economics, and temporal wisdom to craft resolutions. This approach echoes the United Nations but is sculpted for a galactic and temporal stage, with t_3 ensuring that every action aligns with the universe's intricate design. Centrum's neutrality will shield it from becoming a domineering hub like the imperial capitals of old, fostering a network alive with the exploratory spirit of Homo explorans.

The scale of this cosmic expansion favors the Oikumene over an empire. Within 100 light-years, over 2,000 potentially habitable planets—including Proxima Centauri b, TRAPPIST-1's worlds, and TOI-700 d—beckon with diverse landscapes that will shape distinct societies. A high-gravity world might nurture a culture of stoic endurance, while a water-rich planet could inspire aquatic traditions. Time travel will open doors to Earth's past and future, allowing humanity to converse with ancient philosophers or glean wisdom from distant centuries. Navigation across alternate timelines, steadied by t_3, will unveil billions of realities—some radiant with harmony, others shadowed by strife. An empire enforcing uniform laws from Earth would struggle to govern such diversity, particularly across timelines where t_3's oversight ensures causal integrity. The Oikumene's autonomy allows each planet and timeline to craft its own path, while Centrum's diplomacy resolves disputes with grace, allowing the restless curiosity of Homo explorans to flourish.

Autonomy is the Oikumene's cornerstone, enabling societies to thrive amid diverse environments and histories. A planet scarce in resources might embrace a communal ethos, while a timeline rich in innovation could champion progress. This mirrors the adaptability of 17th-century New England colonies, which flourished by tailoring governance to local needs, unlike Spain's rigid dominions. Advanced communication—through quantum entanglement or timeline-linked relays—will allow instant sharing of knowledge, fostering collaboration without a central overlord. The Princeps on Centrum will guide these exchanges, offering aid in crises, such as a planet's ecological collapse or a timeline's unrest, without dictating terms. The t_3 axis will ensure these actions respect causal boundaries, allowing Homo explorans to explore myriad realities while preserving the universe's delicate balance.

Economically, the Oikumene will weave a tapestry of prosperity free from imperial exploitation. The Age of Exploration concentrated wealth in Europe, leaving inequality in its wake. The Oikumene will enable planets to specialize—lithium from exomoons, helium-3 for energy, or biotechnologies from advanced timelines—creating a vibrant galactic and temporal economy. Voluntary trade pacts, mediated by the Princeps, will prevent monopolies, unlike Britain's mercantilist Navigation Acts. Freighters crossing star systems in months and knowledge shared across timelines will ensure thriving markets. The t_3 axis will guard against causal disruptions, so that a timeline's history, for example, is not disrupted by an attempt to adopt its technology. This vast economy, fueled by Homo explorans's curiosity, will sustain a population soaring to hundreds of billions, nourished by the bounty of over 2,000 worlds and countless timelines.

Cultural diversity will blossom under the Oikumene, as planets and timelines forge unique identities. A desert world might cultivate nomadic traditions, while a timeline where the arts flourish could inspire transcendent aesthetics. An empire might stifle these voices, but the Oikumene, by granting autonomy, will encourage diversity while fostering exchange. Instant communication will allow artists and thinkers to share ideas across light-years and timelines, igniting a cosmic renaissance. Centrum will host archives and festivals, with the Princeps easing disputes over intellectual property or cultural missteps. The t_3 axis will ensure these exchanges honor causal integrity, preventing timeline disruptions. This cultural flowering, driven by Homo explorans's zeal for discovery, will echo the Hellenistic era's creative syncretism on a cosmic scale, enriching humanity's collective soul.

The Oikumene is uniquely poised to navigate encounters with other intelligent civilizations. With over 2,000 potentially habitable planets within 100 light-years of Earth, life may abound, though intelligent beings are likely rare. An empire might approach alien contact with the aggression of colonial powers in the Americas. The Oikumene, rooted in diplomacy, will foster harmony. The Princeps will orchestrate knowledge-sharing, such as adopting an alien civilization's biotechnology, ensuring mutual gain. The t_3 axis will guide these interactions, preventing temporal ripples from alien technologies or timeline encounters. Should disputes arise over a habitable planet, Centrum's mediators will craft solutions honoring both human and alien dignity, embodying Homo explorans's commitment to ethical exploration.

The risk of conflict with alien civilizations highlights the Oikumene's strengths. Historical empires vied for resources, often trampling indigenous rights. An interstellar empire might see alien worlds as prizes justifying war. The Oikumene's policy of respecting planetary autonomy will allow individual planets to choose diplomacy or neutrality, with the Princeps ensuring peaceful resolutions. Ethical guidelines, perhaps upheld by AI mediators or galactic treaties, will protect alien ecosystems, avoiding colonial missteps. The t_3 axis will ensure these interactions preserve causality, preventing timeline-altering conflicts. The Oikumene's diversity will render humanity resilient, as no single loss will be able to shatter the species, unlike heavily centralized empires, which can prove remarkably fragile in adversity.

The Oikumene will nurture an era of human flourishing, fulfilling Homo explorans's aspirations. By focusing on uninhabited planets and timelines, it will sidestep colonial exploitation. Abundant resources—minerals, energy, and wisdom from advanced timelines—will banish scarcity, ensuring universal access to sustenance, health, and learning. Autonomy will allow planets and timelines to craft tailored societies, from communal havens to meritocratic realms. The Princeps will coordinate knowledge-sharing, spreading innovations across the cosmos. The t_3 axis will ensure these advancements respect causal boundaries, fostering sustainable progress. This cosmic enlightenment, echoing the post-1492 Enlightenment, will enable hundreds of billions to live lives of creativity and purpose, driven by Homo explorans's boundless curiosity.

Colonial empires often bred inequality and revolt, as seen in the American Revolution. An interstellar and inter-temporal empire would magnify these flaws, ignoring local needs and causal complexities. The Oikumene's planets and timelines will govern themselves, joining cooperative networks freely. Centrum's role, confined to diplomacy, will prevent imperial overreach, with the Princeps convening assemblies to negotiate trade, defense, or temporal ethics. The t_3 axis will ensure decisions honor causality, fostering a balanced civilization.

The practical demands of galactic governance favor the Oikumene. An empire overseeing thousands of planets and billions of timelines would require a sprawling bureaucracy. The Oikumene's local governance, with planets and timelines managing their own affairs, will be agile, with Centrum coordinating shared concerns like trade routes or cosmic threats. The Princeps's lean administration, funded by voluntary contributions from planetary governments, will be sustainable, unlike an empire's colossal costs.

The Oikumene's flexibility will temper the risks of humanity's expansion. Colonizing over 2,000 planets and navigating billions of timelines could strain resources or populations. An empire's uniform policies might ignore local limits, just as colonial powers once overexploited resources. The Oikumene will favor local planetary policies tailored to local conditions, ensuring sustainability. The Princeps will facilitate agreements to protect alien ecosystems or timeline integrity, with t_3 ensuring ethical exploration.

An Oikumene of autonomous planets and timelines, united by cooperation and led by the Princeps on Centrum, is the most just and enduring framework for humanity's cosmic future. It will sidestep the pitfalls of empires, fostering innovation, sustainability, and diversity across thousands of worlds and billions of timelines. The Oikumene will nurture a vibrant economy, celebrate cultural richness, and promote human flourishing. In encounters with alien civilizations and other timelines, the Princeps's diplomacy will ensure peaceful coexistence, embodying Homo explorans's exploratory ethos without needless conflicts. As humanity sails into the ever-widening universe, the Oikumene offers a model of freedom, collaboration, and moral responsibility, ensuring our species thrives among the stars and across the vastness of time.

Chapter XXIII

Exploration of the Universe as a Path to Theosis

Humanity has always been drawn to the stars, our eyes lifted to the heavens with a mixture of awe and longing. This urge to explore the unknown, to venture beyond the horizons of our world, is more than a quest for knowledge—it is a fundamental expression of our being, a spark of divine curiosity embedded in our souls. The vastness of the cosmos, with its shimmering galaxies and silent voids, seems to beckon us, whispering questions that have echoed through the ages: What lies beyond? Who are we in the grand design of creation? In the theology of Saint Gregory Palamas, a 14th-century Hesychast mystic, we find a profound lens through which to view this cosmic journey. His teachings on God's Divine Energies and the pursuit of theosis—union with the Divine—illuminate our exploration of the universe as a sacred path, one that mirrors the Creator's intelligence and draws us ever closer to His eternal beauty. Through the lens of six-dimensional spacetime, a mathematical vision of a universe teeming with hundreds of billions of timelines, we see not only a physical cosmos but a reflection of God's boundless intelligence and creativity. As we stand on the cusp of technologies that may one day enable faster-than-light travel, time travel, and journeys to alternate timelines, our exploration becomes a dual pilgrimage—outward into the exquisite vastness of creation and inward toward the divine presence within, guiding us toward the ultimate aim of theosis.

The impulse to explore is woven into the fabric of human existence. From the earliest hominids who wandered beyond the safety of their caves to the Polynesian navigators who charted the Pacific by starlight, humanity has been driven by a restless curiosity. This drive is not merely practical—a search for resources or territory—but deeply existential, rooted in our capacity to imagine what lies beyond the known. The ancient Greeks, with their myths of Icarus soaring toward the sun, captured this yearning. Alexander the Great marched from Greece to India, driven by a desire to conquer the edges of the world. In the Renaissance, Galileo turned his telescope to the heavens, defying dogma to reveal a cosmos far grander than the Church had imagined. In the 20th century, humanity took its first steps beyond Earth, landing on the moon and sending probes to Mars and the outer reaches of the solar system. Today, the discovery of exoplanets like Proxima Centauri b and Kepler-442b fuels dreams of interstellar colonies, of worlds where humanity might one day walk. Each of these milestones reflects a fundamental trait of our consciousness: the need to transcend boundaries, to touch the unknown, and to understand our place in the grand design.

This exploratory spirit is not a trivial pursuit but a hallmark of consciousness itself. To be conscious is to reflect, to question, and to seek meaning. Our minds, with their ability to imagine possibilities beyond the immediate, resonate with the vastness of the cosmos. When we gaze at the stars, we are not merely observing distant points of light but engaging with the mystery of existence. This resonance suggests a profound connection: by exploring the universe, we are not only cataloging its wonders but seeking to understand the mind of its Creator. The universe, in its infinite complexity, becomes a mirror of divine intelligence, a creation suffused with purpose and beauty.

Saint Gregory Palamas offers a theological framework that deepens this understanding. In his Hesychast teachings, Palamas distinguished between God's essence, which is unknowable, and His Divine Energies, the active manifestations of His love, wisdom, and beauty through which He interacts with creation. These Energies are not separate from God but are God Himself, present in the world yet transcending it. For Palamas, the universe is not a mechanical construct but a living expression of these Energies, a tapestry woven with divine intent. The laws of nature—gravity binding stars in elegant orbits, electromagnetism weaving light across the void, quantum mechanics governing the dance of particles—are not arbitrary but reflections of God's creative will. Each scientific discovery, from the cosmic microwave background to the possibility of liquid water on distant exoplanets, reveals a glimpse of this divine activity, inviting us to contemplate the Creator's mind.

The theory of six-dimensional spacetime, a mathematical model that reimagines the structure of the cosmos, enhances this perspective. In this framework, time is not a passive backdrop but the foundation of reality, with space emerging from its dynamic curvature. This model, which in speculative narratives enables feats like faster-than-light travel to Proxima Centauri b in mere months, suggests a universe of extraordinary complexity, one that encompasses hundreds of billions of timelines branching and intertwining. Such a vision aligns with Palamas' theology, portraying the universe as a profound meditation on God's Divine Energies. The elegance of its equations, describing temporal resonance and spatial contraction, mirrors the harmony and purpose of divine creation. This mathematical beauty is not merely abstract but a reflection of the Creator's intelligence, a cosmic symphony that invites us to participate in its unfolding.

For Palamas, the purpose of human life is theosis, the transformative process of becoming one with God through participation in His Divine Energies. Theosis is not a merging into God's unknowable essence but a journey of growing ever closer to His love, wisdom, and beauty. This process requires active engagement with the Divine, often through prayer, contemplation, and virtuous living. Yet, as we venture into the cosmos, we can see exploration itself as a form of this spiritual pursuit. By studying the universe—its galaxies, its black holes, its potential for life—we encounter the Creator's handiwork, drawing nearer to Him through the act of discovery. Each new insight, whether the detection of gravitational waves or the mapping of exoplanetary atmospheres, is a revelation of divine truth, a step toward the beauty that Palamas saw as central to God's nature.

Theosis, in Palamas' view, is a journey of alignment, where the human soul attunes itself to God's Energies. He taught that God resides within the heart of every person, a divine presence that connects us to the Creator. This inner presence mirrors the outer universe, suggesting a profound correspondence between the microcosm of human consciousness and the macrocosm of the cosmos. To explore the stars is to explore this inner divine space, to venture inward even as we journey outward. The six-dimensional spacetime model, with its vision of a universe alive with branching timelines, becomes a canvas for this dual exploration. Each voyage—whether a speculative warp-driven mission to Kepler-442b or a probe's journey to Europa's oceans—is a meditation on the Divine Energies, a step toward aligning our consciousness with the Creator's purpose.

This correspondence between the human mind and the cosmos is not coincidental but rooted in the nature of consciousness itself. The six-dimensional model posits that spacetime arises from temporal dynamics, a curvature that shapes reality. Similarly, human consciousness emerges from the intricate patterns of neural processes, a cognitive curvature that enables thought, imagination, and perception. This parallel suggests that we are conscious because the universe is conscious, imbued with a purposeful intelligence by its Creator. In Palamas' theology, the universe is not a lifeless mechanism but a dynamic expression of divine purpose, suffused with God's Energies. The patterns we observe—galactic spirals, orbital resonances, the fine-tuned constants that allow life—reflect a cosmos designed to be known, a creation that resonates with the human mind.

The idea of a conscious universe may seem bold, but it flows naturally from both Palamas' theology and the implications of six-dimensional spacetime. If the universe is a creation of God's Energies, it is alive with possibility, its structure reflecting divine intent. The t_2 dimension of branching timelines, for instance, echoes the human capacity for imagination, where we explore alternate paths and possibilities. The t_3 dimension, enabling smooth transitions between timelines, mirrors our ability to adapt and grow. These temporal dimensions, if harnessed through future technologies like faster-than-light travel or time manipulation, will allow humanity to traverse light-years and explore alternate realities. Yet they also reveal a deeper truth: the universe is designed for exploration, crafted to invite us into a dialogue with its Creator.

Future technologies, such as those enabling faster-than-light travel, time travel, or journeys to alternate timelines, will amplify this dialogue. These advancements, grounded in the mathematics of six-dimensional spacetime, will not only expand our physical reach but deepen our understanding of the universe's divine structure. A warp drive that contracts space and manipulates time, as imagined in speculative missions to distant stars, would allow us to visit worlds like Proxima Centauri b or Kepler-442b, potentially habitable planets that could harbor life. Time travel, by navigating the branching timelines of the t_2 dimension, could offer insights into the universe's past and future. Travel to alternate timelines, enabled by the t_3 dimension's smooth transitions, might reveal realities where different choices shaped history. These technologies, far from being mere tools, would serve as instruments of theosis, enabling us to explore the Creator's handiwork with unprecedented clarity and grow ever closer to His Divine Energies.

This cosmic exploration is a form of active meditation, akin to the Hesychast practices Palamas described. In Hesychasm, prayer and contemplation allow the practitioner to encounter the Divine Light, a manifestation of God's Energies that transforms the soul. Similarly, the study of the universe—its stars, its laws, its temporal dimensions—engages the mind and spirit in a contemplative act. When we calculate the orbits of planets or model the curvature of spacetime, we are not merely performing science but participating in a divine dialogue. Each discovery is a glimpse of God's wisdom, a moment of alignment with His creative purpose. This process, though potentially endless given the universe's vastness, is not futile but asymptotic, drawing us ever closer to the Divine without ever fully attaining it.

The universe's scale, with its hundreds of billions of galaxies and timelines, underscores the magnitude of this journey. Yet its vastness is not a barrier but an invitation, a reflection of God's infinite creativity. Palamas' concept of Divine Law, expressed through the growth of truth and beauty, finds its echo in the universe's expansion, its formation of stars and planets. Human exploration, enabled by technology and guided by curiosity, is a participation in this creative process. Each new world we discover, each new law we uncover, is a revelation of divine truth, a step toward the beauty that Palamas saw as central to God's nature. The pursuit of theosis, in this sense, is not confined to the cloister but extends to the cosmos.

The technologies of the future will amplify this sacred pursuit. Faster-than-light travel, by collapsing the distances between stars, will allow us to encounter new worlds and perhaps new forms of life, each a testament to God's creative Energies. Time travel, by revealing the unfolding of history, will deepen our understanding of divine providence. Travel to alternate timelines, by showing us the myriad possibilities of creation, will reflect the boundless imagination of the Creator. These advancements will not only expand our knowledge but transform our souls, drawing us closer to the divine intelligence that designed the cosmos. They are tools of theosis, enabling us to participate more fully in God's Energies and to grow in love, wisdom, and beauty.

The exploration of the universe is thus a path toward union with the Divine. Palamas taught that God's Energies are present in all creation, inviting us to engage with His handiwork. Our consciousness, arising from the same curvature that shapes the cosmos, is a gift that enables this engagement. By venturing outward into the stars and inward into the heart where God resides, we align ourselves with the Creator's purpose. The six-dimensional spacetime model, with its vision of a universe alive with possibility, reflects this purpose, portraying a cosmos designed to be explored and understood. Each step—whether a mission to a distant exoplanet or a theoretical breakthrough in temporal dynamics—is a step toward theosis, a meditation on the Divine Energies that suffuse creation.

This journey is endless, yet it is filled with meaning. The universe, with its infinite horizons, mirrors the infinite nature of God's Energies. Our exploration, though it may never reach a final destination, is a process of growth, a movement toward ever-greater alignment with divine truth and beauty. As Palamas taught, theosis is not a static achievement but a dynamic becoming, a lifelong and perhaps eternal pursuit of the Divine. The cosmos, as an expression of God's intelligence, invites us into this pursuit, calling us to explore its wonders and to seek the Creator who fashioned it.

In the end, the human urge to explore the universe is a sacred calling, a reflection of the divine image within us. The theology of Saint Gregory Palamas, with its vision of God's Energies and the pursuit of theosis, reveals the cosmos as a mirror of divine intelligence, wisdom, and beauty. The six-dimensional spacetime model, with its elegant mathematics and boundless timelines, deepens this vision, portraying a universe alive with divine purpose. As we develop technologies to traverse light-years, manipulate time, and explore alternate realities, we can choose to draw ever closer to the Creator, participating in His creative work. This cosmic quest, both outward and inward, is a journey toward theosis, a pilgrimage through the exquisite universe that leads us, step by step, toward the heart of God.

Glossary of Key Terms and Concepts

Alcubierre Drive: A theoretical propulsion concept proposed by physicist Miguel Alcubierre in 1994, allowing faster-than-light (FTL) travel without violating Einstein's relativity. It creates a "warp bubble" that contracts spacetime in front of a starship and expands it behind, letting the ship move faster than light while staying still inside the bubble. Think of it like surfing a wave, where the wave (spacetime) moves, not the surfer (ship). In the Odyssey, the TWG generates this bubble, powered by zero point energy.

Bose-Einstein Condensate (BEC): A state of matter where millions of atoms (in the Odyssey, 2 million rubidium atoms) are cooled to near absolute zero (8 nanokelvin, or -459.67°F) and act like a single quantum wave. In the TWG, the BEC, housed in a 6-meter Penning trap, acts like a "temporal lens," amplifying the t_2 (branching possibilities) dimension to strengthen the warp bubble. Imagine it as a choir singing in perfect unison, creating a powerful, unified effect.

Casimir Cavity Array: A collection of 1 million tiny tungsten-carbide spheres (10 microns wide, smaller than dust) in the ZPE reactor, spaced 1 nanometer apart to harness the Casimir effect. This effect creates negative energy density by limiting quantum vacuum fluctuations between the spheres. In the Odyssey, the array produces 1,000 joules per cubic meter of negative energy to power the TWG's warp bubble. Think of it as squeezing a sponge to extract water, but here, the sponge is empty space, and the water is energy.

Casimir Effect: A quantum physics phenomenon where two objects placed very close together (e.g., 1 nanometer) restrict invisible energy waves in the vacuum, creating a slight "negative energy" effect, where energy is less than zero. This negative energy is crucial for the Alcubierre Drive's warp bubble. In the Odyssey, the Casimir cavity array uses this effect to generate energy. Imagine it like calming a noisy room by closing the doors, leaving a quieter (negative) energy inside.

Cryocooler: A high-tech refrigerator that cools systems to extremely low temperatures, like 3.5 Kelvin (-454°F) for the TWG and ZPE reactor. It uses liquid helium and consumes power (e.g., 200 kW for the TWG, 250 kW for the ZPE reactor) to keep components stable, preventing heat from disrupting delicate processes like laser coherence or quantum dot performance. Think of it as an ultra-cold freezer keeping ice cream solid in a scorching desert.

Hawking Radiation: A theoretical type of radiation predicted to be emitted by black holes, caused by quantum effects near their edges. In the Odyssey, uncontrolled resonance in the ZPE reactor's Casimir cavities could produce similar radiation, posing a risk to the crew. Piezoelectric graphene dampers prevent this by absorbing vibrations. Imagine it as a dangerous hum from a machine that needs to be silenced to avoid harm.

Kletetschka's Six-Dimensional Spacetime Theory: A 2025 theory by Dr. Gunther Kletetschka, published in Reports on Advances in Physical Sciences, proposing that spacetime has three spatial dimensions (x, y, z) and three temporal dimensions: t_1 (linear time, like days passing), t_2 (branching possibilities, like different story endings), and t_3 (transitions linking those possibilities). In the Odyssey, the TWG uses this theory to manipulate t_2 and t_3 with laser pulses, creating the warp bubble. Think of it as a three-lane highway of time, where the ship navigates side lanes to bend space.

Negative Energy Density: A state where energy is less than zero, unlike normal matter (e.g., a 1-kg object has 90,000 trillion joules of positive energy). Required for the Alcubierre Drive's warp bubble, it's produced in the Odyssey by the ZPE reactor's Casimir cavities, amplified by t_2/t_3 resonance. Imagine it as a "debt" of energy that lets you borrow space to move faster.

Penning Trap: A device using electric and magnetic fields to hold charged particles, like 0.1 grams of positrons in the ZPE reactor or 2 million rubidium atoms in the TWG's BEC. In the Odyssey, it operates in an ultra-high vacuum (10^{-12} torr) to prevent particles from hitting matter and causing explosions (annihilation). Think of it as a magnetic cage keeping wild animals safely contained.

Petawatt: A unit of power equal to 1 million billion watts (10^{15} watts), far more than a household light bulb (60 watts). The Odyssey's TWG requires 9.6 petawatts to start the warp bubble and 0.8 petawatts to keep it going, while the ZPE reactor's lasers deliver 1.5 petawatts per pulse. Imagine it as the energy to light up an entire country in a split second.

Petahertz: A frequency unit meaning 1 quadrillion cycles per second (10^{15} Hz), far beyond visible light's frequencies (hundreds of terahertz). In the Odyssey, the TWG and ZPE reactor use 1.5-petahertz laser pulses to resonate with t_2 and t_3, shaping the warp bubble. Think of it as a super-fast drumbeat, vibrating billions of times faster than a hummingbird's wings.

Piezoelectric Graphene Dampers: Thin layers of graphene, a single sheet of carbon atoms stronger than steel, that absorb tiny vibrations (0.01 mm) in the ZPE reactor. They work using the piezoelectric effect, where pressure creates electricity, to stabilize the Casimir cavity array and prevent dangerous resonance. Imagine them as shock absorbers on a car, smoothing out bumps to keep the ride steady.

Positrons: Particles identical to electrons but with a positive charge, used in the ZPE reactor's temporal field modulator (0.1 grams). They oscillate at 1.5 petahertz to align energy with the TWG, but if they touch regular matter, they annihilate, releasing energy like a tiny explosion. In the Odyssey, a Penning trap keeps them safe. Think of them as fiery sparks in a secure glass jar.

Quantum Coherence: The ability of a system, like laser light or quantum dots, to stay perfectly synchronized, with waves aligned like soldiers marching in step. In the Odyssey, the TWG and ZPE reactor's lasers achieve 99.999% coherence to ensure precise energy delivery. A 0.01% loss could wobble the warp bubble, like a choir singing slightly off-key.

Quantum Dots: Tiny crystals (4-6 nanometers, smaller than a virus) made of materials like indium arsenide, gallium arsenide, and cadmium selenide. In the Odyssey, they're embedded in the TWG's coils and ZPE reactor's Casimir cavities to control laser light with extreme precision, matching t_2 and t_3 dimensions. Think of them as tiny light switches, directing energy exactly where it's needed.

Quantum Resonance Amplifier: A component of the ZPE reactor using 24 ytterbium-doped lasers (1.5 petawatts, 1.5 petahertz) to boost the Casimir cavities' negative energy 10,000 times. It resonates with t_2 and t_3 to amplify vacuum fluctuations, like turning up the volume on a quiet radio to make it loud enough for the TWG's warp bubble.

Superconducting Solenoid: A coil of wire (niobium-tin in the ZPE reactor, niobium-titanium in the TWG) that conducts electricity without resistance when cooled to 3.5 Kelvin. It generates strong magnetic fields (18 Tesla for the ZPE reactor, 15 Tesla for the TWG) to stabilize energy systems. Imagine it as a powerful magnet guiding energy like a river through a channel.

Temporal Field Modulator: A ZPE reactor component that shapes vacuum energy for the TWG, using an 18-Tesla superconducting solenoid and a Penning trap with 0.1 grams of positrons. It aligns energy pulses every 10 microseconds, guided by laser interferometers, to keep the warp bubble stable. Think of it as a conductor ensuring an orchestra plays in perfect harmony.

Temporal Wave Generator (TWG): The Odyssey's main propulsion engine, an 18-meter toroidal ring on Deck 4, using 1.5-petahertz lasers, quantum dots, and a BEC to create a 50-meter warp bubble. It manipulates t_2 and t_3 dimensions via Kletetschka's Maxwell equation to bend spacetime at 50c. Imagine it as a cosmic sculptor, molding time to move the ship faster than light.

Timeline Detector: A sensor with 10^{-18} second precision that monitors the t_2 dimension to prevent timeline paradoxes (e.g., getting stuck in an alternate reality). In the Odyssey, it works with interferometers and Curva to stabilize the warp bubble. Think of it as a cosmic GPS, ensuring the ship stays on the right path through time.

Tungsten-Carbide Spheres: Tiny (10-micron) spheres in the ZPE reactor's Casimir array, made of a super-strong material to withstand quantum pressures. Coated with quantum dots, they confine vacuum fluctuations to produce negative energy. Imagine them as sturdy buckets collecting water from a faint trickle.

Ultra-High Vacuum (10^{-12} torr): An environment with almost no particles, emptier than deep space, used in the ZPE reactor and TWG to prevent contamination. It's maintained by cryopumps to protect positrons and BEC atoms. Think of it as a perfectly clean room where even a speck of dust could ruin everything.

Warp Bubble: A 50-meter sphere around the Odyssey, created by the TWG, that contracts spacetime in front and expands it behind, enabling 50c travel (15 billion m/s). It allows FTL without moving fast locally, like riding an escalator that moves space itself.

Zero Point Energy (ZPE) Reactor: The Odyssey's power source, a 4-meter-long, 1.5-meter-wide cylinder on Deck 4, producing 5×10^{16} joules per pulse (12 megatons TNT) via Casimir cavities, lasers, and a positron modulator. It harnesses quantum vacuum energy, replacing antimatter, for sustainable FTL travel. Imagine it as a cosmic battery tapping the universe's hidden energy.

Addendum

Estimating the TSWG's Energy Parameters

In the grand tapestry of the cosmos, where stars weave their tales across eons, the notion of traversing spacetime faster than light captivates the human imagination. The Temporal-Spatial Wave Generator (TSWG), as envisioned in the speculative narrative of interstellar and temporal exploration, harnesses the power of light to sculpt a 50-meter Alcubierre warp bubble, enabling a starship to journey 4.24 light-years to Proxima Centauri b in a mere 31 days—a velocity 50 times that of light itself. At the heart of this audacious feat lie the TSWG's lasers, pulsing at 1.5 petahertz with a staggering 1.5 petawatts of power. These parameters, though speculative, are not mere whims; they are educated conjectures, elegantly rooted in the physics of gamma rays and high-energy lasers, poised to bend the very fabric of spacetime. This essay explores why these values—1.5 petahertz and 1.5 petawatts—are necessary and sufficient to conjure the warp bubble, drawing on real-world science to illuminate their plausibility.

The Alcubierre Drive, a theoretical marvel proposed by physicist Miguel Alcubierre, offers a vision of faster-than-light travel that respects Einstein's relativity. It envisions a bubble of spacetime that contracts ahead of a starship and expands behind, propelling it at superluminal speeds while the vessel remains serene in a pocket of flat spacetime. This elegant solution demands a negative energy density—an exotic state that defies conventional matter—to sculpt the requisite curvature. In the TSWG's imaginative framework, this is achieved not with unattainable substances but through the precise dance of laser light, orchestrated by quantum dot arrays, which resonate with a six-dimensional spacetime model to mimic this negative energy. The lasers' output, pulsing at 1.5 petahertz and delivering 1.5 petawatts, is the linchpin of this cosmic choreography, and their necessity is illuminated by the physics of light and energy.

The frequency of 1.5 petahertz, a beacon in the gamma-ray spectrum, is essential for its ability to engage the quantum vacuum with exquisite precision. At this frequency, each photon carries an energy of approximately 6.2 million electron volts (MeV), calculated as \(E = h\nu\), where Planck's constant meets the rapid oscillation of 1.5 quadrillion cycles per second. In the real world, gamma rays of such energy are the heralds of nuclear reactions and cosmic phenomena, wielded in facilities like CERN to probe the subatomic realm or observed in the bursts of distant supernovae. These high-energy photons are capable of interacting with the quantum vacuum, stirring virtual particles that flicker in and out of existence. The Alcubierre warp bubble requires a manipulation of spacetime that mimics negative energy, a feat theoretically linked to quantum effects like the Casimir effect, where virtual particles generate measurable forces. The TSWG's quantum dots—tiny, 4-to-6-nanometer crystals of indium arsenide, gallium arsenide, and cadmium selenide—act as maestros, absorbing and re-emitting these 1.5-petahertz pulses with 99.999% precision. This coherence ensures that the light's energy aligns perfectly with the electron transitions needed to resonate with the speculative t_2 and t_3 dimensions, creating the spatial contractions and expansions of the 50-meter bubble.

The sufficiency of 1.5 petahertz rests on its alignment with the energy scales of real-world gamma-ray applications. In laboratories, gamma rays near this frequency drive particle creation and probe quantum field effects, suggesting their potential to manipulate the vacuum in ways that could, in a speculative leap, produce the negative energy-like effects required for the Alcubierre metric. The TSWG's quantum dots, functioning as artificial atoms, confine electrons to precise energy levels that match these 6.2 MeV photons, ensuring the pulses are sharp and stable. While the Alcubierre metric does not prescribe a specific frequency, the choice of 1.5 petahertz is a reasoned extrapolation, drawing on the real-world efficacy of gamma rays to interact with fundamental quantum processes. This frequency is deemed sufficient to sustain the warp bubble's geometry, enabling the starship to traverse vast distances at 50 times the speed of light, as the essay's narrative assumes a stable bubble enveloping the 150-meter Odyssey.

The power of 1.5 petawatts per laser, totaling 18 petawatts across 12 lasers, is equally vital, providing the immense energy flux needed to drive this cosmic endeavor. In the real world, petawatt lasers illuminate the frontiers of physics. The National Ignition Facility (NIF), for instance, achieves bursts of 1.8 petawatts by delivering 1.8 megajoules in a fleeting billionth of a second, powering experiments in fusion energy. The TSWG's lasers, each pulsing at 1.5 petawatts for 0.3 seconds, contribute approximately 6.48×10^{16} joules to the system's 5×10^{17} joule requirement, a scale that dwarfs earthly power plants but aligns with the speculative ambition of warping spacetime. This energy, distributed across the TSWG's 22-meter toroidal structure, yields a flux of roughly 47 trillion watts per square meter, sufficient to excite the quantum dots and the Bose-Einstein condensate (BEC), which amplifies the pulses to sculpt the warp bubble's curvature.

The necessity of 1.5 petawatts arises from the need to deliver this colossal energy in brief, controlled bursts, ensuring the quantum dots emit coherent light and the BEC generates the rotational quantum field for spacetime manipulation. Real-world lasers, like those at NIF or the Extreme Light Infrastructure, demonstrate that petawatt-scale power can focus energy densities to extreme levels, capable of ionizing matter or probing quantum effects. The TSWG's graphene amplifiers, leveraging graphene's extraordinary conductivity, stabilize these high-power pulses, preventing thermal or phase disruptions. The power level is sufficient, as it supports the essay's claim of a 50-meter bubble, with the energy flux driving the quantum interactions needed to mimic negative energy. The Alcubierre metric's original energy estimate—on the order of 10^{64} joules—is infeasible, but the essay's six-dimensional model posits a drastic reduction, making 18 petawatts a plausible threshold for initiating the bubble, as evidenced by the manageable 3 megawatts of waste heat handled by advanced cooling systems.

In conclusion, the 1.5-petahertz frequency and 1.5-petawatt power of the TSWG's lasers are elegant, educated conjectures, crafted to harmonize with the Alcubierre Drive's theoretical demands and real-world physics. The frequency, rooted in the gamma-ray energies that probe the quantum vacuum, is poised to excite the TSWG's quantum dots, creating the precise light needed to warp spacetime. The power, echoing the might of cutting-edge petawatt lasers, delivers the energy flux to sustain this cosmic alchemy, rendering the 50-meter warp bubble a reality within the essay's visionary framework. While speculative, these parameters dance on the edge of possibility, their sufficiency inferred from their alignment with the energy scales of real-world gamma rays and lasers, illuminating a path where humanity might one day sail the stars.